投考公務員
能力傾向測試
解題天書

第三版

U0130818

試題王 詳盡分析

280條模擬試題，密集特訓

詳盡題型練習及模擬試卷

解題分析演繹推理

投考公務員必備天書

文化會社

Fong Sir 著

序言

公務員綜合招聘考試（Common Recruitment Examination，又稱CRE）是特區政府聘請大部份公務員的第一關，由於通過CRE是應徵學位或專業程度公務員職位的基本條件，故應考人數年年遞增，每次均達數萬人。

政府聘請公務員，一般要求申請人需於CRE中的個別試卷中取得相當的成績，作為衡量申請人語文及推理能力的客觀標準。雖然公務員事務局表示，個別部門可以因應需要而接受未達標的申請，但申請人必須在指定日期參加CRE，並符合招聘廣告內列出的基本入職要求才會獲聘。故此，有意投考公務員空缺者，還是及早應考為上。

由於CRE的試卷並不會公開，令不少有意投身公務員行列人士難以提前做好準備工作，錯失成為公務員的機會。本書根據CRE最常出的題型撰寫。書中不但詳列各類題型的解題方法，更輔以練習、解説及模擬試卷，內容由淺入深，助你提升答題速度，而題目數量多達280條，絕對是有意投考公務員空缺者的必讀之選。

本書的精彩內容如下：

- 收錄多達280條題目，是備戰能力試的操卷首選。

- 深入剖析5大題型的必用概念：中文演繹推理、英文演繹推理、數據運用、數字推理及圖表分析。透過Concept teaching的方法，讓你即使面對新題種，亦能應付自如。

- 精選2份模擬試卷，讓你熟習考試模式及出題方向。通過反覆練習，助你掌握答題技巧，提升答題速度。

- 試題講解詳盡，方便考生藉此檢視自己的強項及弱點，提升學習成效。

目錄

序 　　　　2

Chapter 1 試題練習

I. 演繹推理 　6
矛盾關係的推理 　8
三段論 　11
條件關係推理 　13
歸納推理 　17
削弱一個論證 　20
尋找題幹成立的大前提 　25
解釋 　26
引出結論 　28
評價一個論證 　30
排除法 　31
試題練習：演繹推理 Q & A 　33

II. Verbal Reasoning（English） 　67
解題步驟 (1)：定位與判斷 　68
解題步驟 (2)：「True」的特點 　69
解題步驟 (3)：「False」的特點 　71
解題步驟 (4)：「Can't tell」的特點 　77
試題練習：Verbal Reasoning Q & A 　80

III. Data Sufficiency Test 　87
Data Sufficiency Test 四個注意地方 　88
試題練習：Data Sufficiency Test Q&A 　90

IV. Numerical Reasoning 　114
試題練習：Numerical Reasoning Q&A 　115

V. Interpretation of Tables and Graphs 　120
試題練習：Interpretation of Tables and Graphs Q&A 　121

Chapter 2 模擬試題

Mock Paper 1 　154

Mock Paper 2 　181

Chapter 3 常見問題 　206

試題練習

I. 演繹推理

考生須根據短文的內容，選出一個或一組推論。考生須假定短文的內容都是正確的，憑邏輯推論出最正確的答案。

　　演繹推理旨在考察考生的邏輯推理能力：在每道題中給出一段陳述，這段陳述被假設是正確的，不容置疑的。要求考生根據這段陳述，選擇一個備選答案。正確的答案應與給所的陳述相符合，應不需要任何附加說明即可以從陳述中直接推出。由於在演繹推理中，前提與結論之間有必然性的關聯，結論不能超出前提所斷定的範圍。因此，在解答此種試題時，必須緊扣題目部分陳述的內容，正確答案應與所給的陳述相符。

　　必須注意的是：此類試題的備選答案具有很強的迷惑性，即各個選項幾乎都是有道理的，但有道理並不等於與這段陳述直接相關。正確的答案應與陳述直接有關，即從陳述中直接推出。還有一點必須注意，試題中所給的陳述有的合乎常理，有的可能不太合乎常理。

　　考生尤其要特別注意的是：這段陳述被假設是正確的，不容置疑的。考生不能對試題所陳述的事實的正誤提出懷疑，也不能自作聰明地以自己具備的這方面的知識進行推理得出答案，而完全忽視試題中所陳述的事實。

矛盾關係的推理

【定義】

「矛盾關係」是指兩個語句或命題之間不能同真（必有一假），也不能同假（必有一真）。不能同真，就是說當其中一個命題真時，另一個命題必假；不能同假，就是說當其中一個命題假時，另一個命題必真。

矛盾關係一般以三種形式出現：

1.「A」和「-A」

2.「所有」和「有些」

3.AaB 和（A 且 -B）

拆解矛盾關係解題可分三步走：

1. 找出矛盾關係

2. 撇開矛盾關係

3. 根據已知確定信息推理

下面通過幾道例題，講解矛盾關係在解題中的應用。

【例題 1】

莎士比亞在《威尼斯商人》中，寫富家少女鮑細婭品貌雙全，貴族子弟、公子王孫紛紛向她求婚。鮑細婭按照其父遺囑，由求婚者猜盒訂婚。鮑細婭有金、銀、鉛三個盒子，分別刻有三句話，其中只有一個盒子，放有鮑細婭的相片。

求婚者要通過這三句話，去猜鮑細婭的相片放在哪個盒子裡，就嫁給誰。三個盒子上刻的三句話分別是：

1. 金盒子:「相片不在此盒中」

2. 銀盒子:「相片在鉛盒中」

3. 鉛盒子:「相片不在此盒中」

鮑細婭告訴求婚者,上述三句話中,最多只有一句是真的。

如果你是其中一位求婚者,你認為鮑細婭的相片是放在哪個盒子裡?

    A. 金盒子

    B. 銀盒子

    C. 鉛盒子

    D. 要麼金盒子,要麼銀盒子

    E. 不能確定

【答案及解釋】

A. 推斷:(1)真,(2)真,(3)必假;

(1)真,(2)假,(3)必真;

(1)假,則(2)假,(3)真;

答案:綜合以上所述,且題中說最多只有一句是真的,所以相片在金盒子裡。

B. 另一推斷方法:

(2)真,(3)必假,(1)必真;

(2)假,(3)必真,(1)可能真,也可能假。但題中說只有一句是真,於是就排除了(2)假,(3)真,(1)真的可能性了,即應該是:(2)假,(3)真,(1)假。

答案:相片放在金盒子裡。

C. 用簡單的邏輯形式作假設:

假設金盒子:相片不在此盒中為A;銀盒子:相片在鉛盒中為B;鉛盒子:相片不在此盒中為C。

1. A為真，B和C為假。那麼相片肯定不在金盒子中，則必定在銀盒子或鉛盒子中。但是B和C有互相矛盾，B說相片在鉛盒子中為假，那麼相片肯定不在鉛盒子中，而C又說相片不在鉛盒子中為假，即相片肯定在鉛盒子中。所以，這個答案不存在。

2. B為真，A和C為假。A說相片不在此盒中為假，即相片在A即金盒子中；而C又說相片不在此盒中為假，即相片在C即鉛盒子中。兩者相互矛盾，所以結果不存在。

3. C為真，A和B為假。A說相片不在此盒中為假，即相片在A即金盒子中；B說相片在鉛盒子中為假，即相片不在鉛盒子中；C說相片不在此盒中為真，與A和B都沒有衝突。所以，相片在金盒子中。

4. A、B、C都為假。

根據(2)的推論，A和C不可以同假，所以這種情況的結果也不存在。綜上所述，相片在金盒子中。

【例題2】

某珠寶店失竊，甲、乙、丙、丁四人涉嫌被拘審。四人的口供如下：

甲：案犯是丙。

乙：丁是罪犯。

丙：如果我作案，那麼丁是主犯。

丁：作案的不是我。

四個口供中只有一個是假的。

如果上述斷定為真，那麼以下哪項是真的？

    A. 說假話的是甲，作案的是乙。

    B. 說假話的是丁，作案的是丙和丁。

    C. 說假話的是乙，作案的是丙。

    D. 說假話的是丙，作案的是丙。

    E. 說假話的是甲，作案的是甲。

【答案及解釋】

乙和丁的口供矛盾，根據矛盾律，其中必有一假。四人口供中只有一假，所以，甲和丙口供是真的。甲口供真，作案者為丙，加上丙的口供，根據充分條件假言推理肯定前件式，丁也是作案者，由此也斷定乙和丁中，丁講假話。

三段論

【定義】

「三段論」是指由三個命題構成的推理。具體說來，三段論是由包含著一個共同因素（邏輯中介）的兩個命題，而推出一個新命題的推理。例如：

1. 所有闊葉植物都是落葉的

2. 所有葡萄樹都是闊葉植物

3. 所以，所有葡萄樹都是落葉的

上述推理中的共同因素就是「闊葉植物」

進行「三段論」推理，關鍵就是要看這個共同因素能否把兩個前提連接起來推出結論。如果連接不起來，則三段論就是錯誤的。例如：

1. 英雄難過美人關

2. 我難過美人關

3. 所以，我是英雄

上述推理的錯誤就是「難過美人關」這個共同因素沒有能夠把兩個前提必然地連接起來。因為很可能英雄是難過美人關的一種人，但我卻是難過美人關的另一種人。

【例題】

在某住宅小區的居民中，大多數中老年教師都買了人壽保險，所有買了四房以上住宅的居民都買了財產保險。而所有買了人壽保險的都沒有買財產保險。

如果上述斷定是真的，以下哪項關於該小區居民的斷定必定是真的？

1. 有中老年教師買了四房以上的住宅

2. 有中老年教師沒有買財產保險

3. 買了四房以上住宅的居民都沒有購買人壽保險

    A. 1、2和3

    B. 僅1和2

    C. 僅2和3

    D. 僅1和3

    E. 僅2

【答案及解釋】

大多數中老年教師買了人壽保險，而所有買了人壽保險的居民都沒有買財產保險，所以大多數中老年教師沒有買財產保險，這是(2)。買了四房以上住宅的居民都買了財產保險，而所有買了人壽保險的居民都沒有買財產保險，所以，買了四房以上住宅的居民都沒有買人壽保險(否則就矛盾了)，這是(3)。中老年教師和四房以上住房之間沒有建立因果關係，推不出(1)來。

條件關係推理

【定義】

「條件關係」的表現形式主要有三種，即「充分條件關係」和「必要條件關係」：

1. 充分條件：有之必然，無之未必不然

2. 必要條件：無之必不然，有之未必然

與上述條件關係相對應，就有相應的條件關係命題，即充分條件命題和必要條件命題。例如：

1. 如果天下雨，那麼地濕。

2. 只有年滿18歲，才有選舉權。

在條件命題中，表示事物情況存在的條件的部分稱為「前件」，表示依賴條件而存在的部分稱為「後件」。

充分條件命題只要在前件為真，並且後件為假時才是假的，其他情況下都是真的。

在我們的日常溝通中，往往都會用到「如果…就…」、「有…就有…」、「倘若…就…」、「哪裡有…哪裡就有…」、「一旦…就」、「假若…則…」、「只要…就…」等連結詞，這些都能表達充分條件命題。

必要條件命題只有在前件為假，並且後件為真時才是假的，其他情況下都是真的。

在日常語言中，「沒有…就沒有…」、「不…不…」、「除非…不…」、「除非…才…」、「除非…否則不…」、「如果不…那麼不…」等聯結詞都能表達必要條件命題。

充分條件和必要條件之間存在著密切的關係，這就是：

如果 p 是 q 的充分條件，那麼 q 就是 p 的必要條件；

如果 p 是 q 的必要條件，那麼 q 就是 p 的充分條件。

充分條件推理有兩個有效的推理形式：

1. 肯定前件式

A. 如果 p，那麼 q

B. 因 p

C. 所以是 q

2. 否定後件式

A. 如果 p，那麼 q

B. 因非 q

C. 所以非 p

必要條件推理有兩個有效的推理形式：

1. 否定前件式：

A. 只有 p，才 q

B. 因非 p

C. 所以非 q

2. 肯定後件式：

A. 只有 p，才 q

B. 因 q

C. 所以 p

【例題1】

1.「只有認識錯誤，才能改正錯誤。」

以下各項都準確表達了上述斷定的含義，除了：

　　A. 除非認識錯誤，否則不能改正錯誤。

　　B. 如果不認識錯誤，那麼不能改正錯誤。

　　C. 如果改正錯誤，説明已經認識了錯誤。

　　D. 認識錯誤，是改正錯誤的必不可少的條件。

　　E. 只要認識錯誤，就一定改正錯誤。

【答案及解釋】

題幹講的是，「認識錯誤」是「改正錯誤」的必要條件。

選項A的聯結詞是「除非…否則不…」。

選項B的聯結詞是「如果不…那麼不…」，與題幹的意思完全一致。選項C説「改正錯誤」是「認識錯誤」的充分條件。

選項D強調「認識錯誤」是「改正錯誤」的必要條件，也與題幹的意思是一致的。

選項E則認為，「認識錯誤」是「改正錯誤」的充分條件，顯然不能表達題幹的意思。所以，正確答案是E。這裡需要注意，「只要……就……」與「只有…才…」不同，前者表達充分條件，後者表達必要條件。

【例題2】

世界級的馬拉松選手每天跑步都不超過6小時。一名選手每天跑步超過6小時，因此他不是一名世界級馬拉松選手。

以下哪項與上文推理形式相同？

　　A. 跳遠運動員每天早晨跑步。如果某人早晨跑步，那麼他是跳遠運動員。

　　B. 如果每日只睡4小時，對身體不利。研究表明，最有價值的睡眠都發生在入睡後的第5個小時。

C. 家長和小孩玩遊戲時，小孩更高興。因此，家長應該多玩遊戲。

D. 如果某汽車早晨能起動，那麼晚上也能起動。我們的車早晨通常能起動，同樣，它晚上通常也能起動。

E. 油漆三小時內都不會乾掉。某塗料在三小時內乾了，所以它不是油漆。

【答案及解釋】

題幹中的第一句話是「全稱否定命題」（即斷定一類對象中全體對象，並不具有某種性質的命題），但我們可以將它作為一個充分條件的「假言命題」（「如果 A 則 B」的複合命題）來看待，即「如果一個人是世界級的馬拉松選手，剛他每天跑步不超過 6 小時。」

選項 E：第一句話也可以看成是一個充分條件的假言命題，即「如果一種塗料是油漆，那麼它在三小時內不會乾掉。」

這樣，題幹和選項 D 的推理形式都屬於「充分條件」假言推理的「否定後件式」，即：

1. 如果 P 則 Q

2. 非 Q

3. 所以非 P

而其他幾個選項，可作以下分析：

選項 A：屬充分條件的假言推理的「肯定後件式」：

1. 如果是 P 則 Q

2. 是 P

3. 所以 P

（例如「如果他生病了，則他不來上班；他不來上班；所以他生病了。」）

選項 B：屬於歸納推理。

選項 C：屬充分條件的假言推理的肯定前件式。

1. 如果 P 則 Q

2. 是 P

3. 所以 Q

所以正確答案是 E。

歸納推理

【定義】

「歸納法」或「歸納推理」，有時叫做「歸納邏輯」，是論證的前提支持結論但不確保結論的推理過程。它基於對特殊的代表的有限觀察，把性質或關係歸結到類型；或基於對反覆再現的現象的模式的有限觀察，公式表達規律。

例如，使用歸納法在如下特殊的命題中：

1. 冰是冷的

2. 彈子球在擊打球桿的時候移動

推斷出普遍的命題如：

1. 所有冰都是冷的

2. 所有彈子球都在擊打球桿的時候移動

近年出現了一類只注重推理、不注重真假的結構比較題目，這類題目只管推理的形式是否一致，不在乎推理是否符合邏輯。或者說不在乎推理是真或者是假。我們來看一段經典的語言。

「如果我有一千萬，我就能買一棟房子。我有一千萬嗎？沒有。所以我仍然沒有房子。

如果我有翅膀，我就能飛。我有翅膀嗎？沒有。所以我也沒辦法飛。

如果把整個太平洋的水倒出，也澆不熄我對你愛情的火。整個太平洋的水全部倒得出嗎？不行。所以我並不愛你。」

這幾個推理用形式化的語言表達就是A→B，-A→-B，這個推理在邏輯裡面是否命題推理是不一定成立的。可是其形式是一致的，因此，從結構比較的角度來看，三者的結構是完全一樣的。今天的公務員考試試題，就是要找這種完全一樣的推理形式。

【例題1】

許多上了年紀的北京人都對小時候廟會上看到的各種絕活念念不忘。如今，這些碩果僅存的文化遺產有了一個更為正式的稱呼——民間藝術。然而，隨著社會現代化進程加快，中國民俗文化面臨前所未有的生存危機。城市環境不斷變化，人們興趣愛好快速分流和轉移，加上民間從事藝術的人才逐漸流失，這一切都使民間藝術發展面臨困境。從這段文字可以推出：

    A. 市場化是民間藝術的出路
    B. 民俗文化需要搶救性保護
    C. 城市建設應突出文化特色
    D. 應提高民間從事藝術人才的社會地位

【答案及解釋】

只看選項可以迅速排除C、D，因為C和D中均包含「應」該，太過絕對。再大致看一下題幹，發現「市場化」屬於文中未提概念，因此選B。

【例題2】

政府規定汽車必須裝安全帶的制度是為了減少車禍傷亡，但在安全帶保護下，司機將車開得更快，事故反而增加了。司機有安全帶保護，自身傷亡減少了，而路人傷亡增加了。

這一事實表明：

    A. 政府對實施效果考慮不周的制度往往事與願違
    B. 安全帶制度必須與嚴格限速的制度同時出台
    C. 汽車裝安全帶是通過犧牲路人利益來保護司機的措施
    D. 制度在產生合意結果的同時也會產生不合意的結果

【答案及解釋】

答案：A。因句中有「往往」這一不表示確定意義的特徵詞；同時「制度」前面有限定詞「對實施效果考慮不周的」，故選A。

選項B：句中的「必須」太絕對，同時「嚴格限速的制度」屬於文中未提概念，故排除。

選項C：無法從題幹中推出結論。

選項D：「制度」擴大了「安全帶制度」這一概念的範圍，並且「制度」前面沒有限定詞，太絕對了，排除。

【例題3】

據一項醫學研究發現，每天睡眠不足7小時的人普遍比睡眠時間更長的人胖。這是因為睡眠不足影響了新陳代謝功能，使刺激食欲的荷爾蒙增加，同時使產生飽脹感的荷爾蒙水平降低。

由此不能推出：

A. 肥人的食欲比體重正常的人好

B. 睡眠狀況影響新陳代謝

C. 荷爾蒙水平影響食欲

D. 肥人的荷爾蒙水平可能異於常人

【答案及解釋】

直接看選項，就可以迅速選出正確答案：A，因為選項B和C中有「影響」這一不表示確定意義的特徵詞。

選項D中有「可能」這一不表示確定意義的特徵詞。

「每天睡眠不足7小時的人普遍比睡眠時間更長的人胖」可以推出B選項；「睡眠不足影響了新陳代謝功能，使刺激食欲的荷爾蒙增加，同時使產生飽脹感的荷爾蒙水平降低」可以推出C選項；A、D中的「胖人」都擴大了「每天睡眠不足七小時的人」的範圍，相比較A說的太絕對，D選項中有「可能」這一不表示確定意義的特徵詞，因此選A。

削弱一個論證

【定義】

「削弱一個論證」是指題幹中給出一個完整的論證或者表達某種觀點，要求從備選項中尋找到最能反駁或削弱題幹的選項。此類試題的提問方式一般是：

1.「以下哪項如果為真，最能削弱上述論證？」

2.「以下哪項如果為真，能夠最有力地削弱上述論證的結論？」

3.「以下哪項如果為真，最可能削弱上述推斷？」

解答這種試題時，考生可以直接去尋找最能削弱題幹的選項。所謂「削弱題幹」，也就是要與題幹唱反調。

一般來說，削弱題幹有兩種方式：一是截斷題幹中的論據與論題的邏輯聯繫，又稱為「截斷關係法」；二是弱化題幹中的論據，又稱為「弱化論據法」或「釜底抽薪法」。

截斷關係法：當題幹中所表達的是「如果 p 那麼 q」這樣的充分條件關係時，我就要找到「p 但並不 q」這樣的選項來削弱它；當題幹中所表達的是「只有 p 才 q」這樣的必要條件關係時，我就要找到「非 p 也 q」這樣的選項來削弱它；當題幹中不一定是某種確定的條件關係時，我要尋找的也是與題幹意思唱反調的選項。

弱化論據法：指所要尋找的選項能夠起到將題幹的論據抽掉，或者使題幹中的論據的支持作用減弱，從而使題幹中的論題不成立或者使題幹中的論題得不到充分的論證。

削弱題幹的方式有多種：削弱論點、削弱論據、削弱論證關係。下面講講如何削弱論證關係。削弱論證關係可以從（a）論據與論點沒有聯繫或有差異、（b）因果倒置、（c）另有他因、（d）反對方法等方面考慮：

A. 論據與論點之間沒有聯繫或有差異

如果論據與論點之間沒有聯繫或有差異，那麼根據該論據不必然得出論點，也就很好地起到了削弱作用。

【例題】

在市場經濟條件下，每個商品生產的經營者都是獨立的經濟主體，都有充分的自主權。因此，他們生產什麼，如何生產都由自己說了算。

以下哪項最能削弱上述結論？

A. 商品生產的經營者都是獨立的經濟主體就意味著由自己決定自己的命運。

B. 商品生產經營者享有充分的自主權利這就意味著由自己決定生產什麼。

C. 商品生產經營者必須了解市場行情和消費者的需求等，才能生產出適銷對路的產品。

D. 每個商品生產的經營者雖然是獨立的經濟主體，但是在經營中也要顧及他人的利益。

【答案及解釋】

答案：C

題幹中的結論是「商品生產經營者生產什麼，如何生產都是自己說了算」，得出該結論的依據是「生產經營者都是獨立的經濟主體，有充分的自主權」，要削弱該結論，只要說明還有其他原因影響生產經營者的決策，而 C 項就指出生產什麼樣的產品需要了解市場行情和消費者的需求，因此也就削弱了結論。A、B 兩項不僅不能對上述結論起到削弱作用，反而都有所加強；D 項削弱程度不及 C 項。故答案選 C。

B. 因果倒置

如果某兩類因素A和B緊密相關，題幹就指出A是造成B的原因，那麼要削弱它，就可以說明B才是造成A的原因，這就是所謂的「因果倒置」。

【例題】

相比那些不踢足球的大學生，經常踢足球的大學生的身體普遍健康些。由此可見，足球運動能鍛煉身體，增進身體健康。

以下哪項為真，最能削弱上述論斷？

 A. 大學生踢足球是出於興趣愛好，不是為了鍛煉身體。

 B. 身體不太好的大學生一般不參加激烈的足球運動。

 C. 足球運動有一定的危險性，容易使人受傷。

 D. 研究表明，長跑比踢足球更能達到鍛煉身體的目的。

【答案及解釋】

答案：B

題幹由「經常踢足球的大學生普遍比不踢足球的身體好」作為論據得出「足球運動能鍛煉身體，增進身體健康」的結論。

B項指出，身體不好的大學生一般不參加激烈的足球運動，即題幹的論證是「因果倒置」，有力地削弱了題幹論斷。A、C、D三項均屬於無關項，不能對題幹論斷提出質疑。故答案選B。

C. 另有他因

有些題目的題幹給出一個結論，同時也會給出得出這個結論的一個原因，這時要對題幹進行削弱，只需要說明得出這個結論的不是這個原因或不僅僅是這個原因，而是另有其他原因，也就達到了削弱的目的。如果題幹是以一個事實、研究、發現或一系列數據為前提推出一個解釋上述事實或數據的結論，要削弱這個結論，就可以通過指出有其他可能來解釋該事實，即存在別的因素影響推論。

【例題】

據統計結果顯示，在韓國由於「心理問題」接受心理諮詢的大學生數量大幅增加，以一間大學諮詢中心為例：2003年咨詢案例為1,364宗，到了2006年達到3,485宗，增加了1.5倍。首爾大學的大學生活文化院諮詢人數也是從2004年的191人，增加到了2005年的285人、2006年的320人，呈逐年遞增趨勢。

下列哪個成立，最能嚴重削弱「諮詢案大幅上升的原因是大學生心理問題增多」的觀點？

A. 未來前途和就業方面的精神壓力增大，承受精神痛苦的韓國大學生驟增。

B. 2004年以前韓國的大學生，即使出現心理問題，也基本不接受心理諮詢。

C. 調查表明最近幾年，韓國某大學諮詢中心的諮詢案件反而呈下降趨勢。

D. 從2005年開始，韓國許多大學成立了心理諮詢中心。

【答案及解釋】

答案：B

題幹的觀點是「諮詢案大幅上升的原因是大學生心理問題增多」，要削弱該結論，只要説明諮詢案大幅上升是由其他原因引起的即可。

A項指出有心理問題的韓國大學生人數增多了，這有可能支持「諮詢案大幅上升的原因是大學生心理問題增多」這一觀點。

C項指出韓國某個大學咨詢中心的諮詢案件呈下降趨勢，通過這一特例只能説明該大學的具體問題，不能推出其他結論。

D項是意圖説明2005年以前很多韓國大學生想去心理諮詢，但是沒有諮詢中心，這跟心理問題是否增多沒有必然聯繫。

只有B項最能削弱「諮詢案大幅上升的原因是大學生心理問題增多」這一觀點，因為如果2004年以前大學生出現心理問題不去接受諮詢的話，那麼2004年以後這種狀況得到了改變他們就會去接受心理諮詢，很有可能2004前後有心理問題的大學生數量沒有太大變化，而是這種習慣的改變導致諮詢案大幅上升。因而答案選B。

D. 反對方法

有些削弱型題目的推理可以簡化為「為達到一個目的而提出一個方法」，要削弱題幹，就要指出該方法是不可行的。

【例題】

為了挽救瀕臨滅絕的大熊貓，一種有效的方法是把它們都捕獲到動物園進行人工飼養和繁殖。

以下哪項如果為真，最能對上述結論提出質疑？

    A. 近5年在全世界各動物園中出生的熊貓總數是9隻，而在野生自然環境中出生的熊貓的數字，不可能準確地獲得。

    B. 只有在熊貓生活的自然環境中，才有它們足夠吃的嫩竹，而嫩竹幾乎是熊貓的惟一食物。

    C. 動物學家警告，對野生動物的人工飼養將會改變牠們的某些遺傳特性。

    D. 提出上述觀點的是一個動物園主，他的建議帶有明顯的商業動機。

【答案及解釋】

答案：B

題幹的推理過程是：為了達到挽救大熊貓的目的，可以採取捕獲到動物園進行人工飼養和繁殖的方法。要削弱該結論，只要說明採用這種方法並不能達到目的即可。

B項指出如果人工飼養，則熊貓沒有嫩竹可吃，無法達到挽救大熊貓的目的，削弱了題幹。

A項自然環境中熊貓的確切數量與題幹論證無關。

C項改變遺傳特性並不代表不能挽救；「帶有明顯的商業動機」不等同於該建議不可行。

D項不一定能質疑題幹。

尋找題幹成立的大前提

【定義】

前提型試題是在題幹中給出結論和部分前提，要求從備選項中找到另一部分前提來將推理補充完整的試題。此類試題的題幹中往往給出小前提（即推理的條件）和結論，要求尋找到一個大前提將題幹中的小前提和結論聯結起來，這也就是人們通常說的「搭橋」。此類試題的提問方式一般是：

「上述推論基於以下哪項假設？」

「以下哪項都可能是上述論證所假設的，除了⋯」

「上述陳述隱含著下列哪項前提？」

「上述論斷是建立在以下哪項假設的基礎上？」

【例題】

當前的大學教育在傳授基本技能上是失敗的。有人對若干大公司人事部門負責人進行了一次調查，發現很大一部分大學畢業生的工作人員中，都沒有很好掌握基本的寫作、數量和邏輯技能。

上述論證是以下列哪項為前提的？

    A. 現在的大學裡沒有基本技能方面的課程。

    B. 新到職的人中極少有大學生。

    C. 寫作、數量、邏輯方面的基本技能對勝任工作很重要。

    D. 大公司內新到職的大學畢業生基本代表了當前大學畢業生的水平。

    E. 過去的大學生比現在的大學生接受了更多的基本技能教育。

【答案及解釋】

答案：C

題幹中要從「若干大公司很大一部分大學畢業生的工作人員都沒有很好掌握基本的寫作、數量和邏輯技能」推出「當前的大學教育在傳授基本技能上是失敗的」的結論，還需要假設一個大前提「大公司的大學畢業生基本代表了當前大學畢業生的水平」，即指出前提和結論之間存在本質聯繫。

選項B與題幹不相關，選項A和D都只是講了當前教育中存在的問題，而沒有能夠把這個問題與「大公司的大學畢業生」的情況聯繫起來，故都不能起到保證題幹論證成立的作用。

解釋

【定義】

「解釋型試題」是題幹中給出一個似乎矛盾實際上並不矛盾的現象，要求從備選項中尋找到能夠解釋的選項。此類試題的提問方式一般是：

「以下哪項如果為真，能最好地解釋上面的矛盾？」

能夠解釋題幹的選項顯然是能夠將題幹中的矛盾解釋清楚，即其實並不矛盾，那麼題幹中為何又顯得很矛盾呢？原因可能是還有某方面的細節沒有考慮到。

【例題1】

二氧化硫是造成酸雨的重要原因。某地區飽受酸雨困擾，為改善這一狀況，該地區1至6月累計減排11.8萬噸二氧化硫，同比下降9.1%。根據監測，雖然本地區空氣中的二氧化硫含量降低，但是酸雨的頻率卻上升了7.1%。

以下最能解釋這一現象的是：

A. 該地區空氣中的部分二氧化硫，是從周圍地區飄移過來的。

B. 雖然二氧化硫的排放得到控制，但其效果要經過一段時間才能顯現。

C. 汽車的大量增加加劇了氮氧化物的排放，而氮氧化物也是造成酸雨的重要原因。

D. 盡管二氧化硫的排放總量減少了，但二氧化硫在污染物中所佔的比重沒有變。

【答案及解釋】

答案：C

酸雨增加顯然不是二氧化硫引起的，因此要找其它的原因。而C選項中汽車的大量增加加劇了氮氧化物的排放，而氮氧化物也是造成酸雨的重要原因。故選C。

【例題2】

皮膚中膠原蛋白的含量決定皮膚是否光滑細膩，決定人的皮膚是否年輕。相同年齡的男性和女性皮膚中含有相同量的膠原蛋白，而且女性更善於保養，並能從日常保養中提高皮膚膠原蛋白含量，盡管如此，女性卻比男性更容易衰老。

以下選項能解釋上述矛盾的是：

A. 男性皮膚內膠原蛋白是網狀結構的，而女性是絲狀結構的。

B. 女性維持光滑細膩的皮膚、年輕美貌的容顏需要大量膠原蛋白。

C. 只有蹄筋類食物富含膠原蛋白，但是很難被人體消化吸收。

D. 男性的膠原蛋白幾乎不消耗，而女性代謝需要消耗大量膠原蛋白。

【答案及解釋】

答案：D

選項A雖然指出男性與女性皮膚內膠原蛋白的結構不同，但並未說明其與「女性比男性更容易衰老」的關係；選項B、C都與題幹矛盾無關；D項則合理解釋了矛盾。故選D。

引出結論

【定義】

結論型試題是在題幹中給出前提，要求推出結論的試題。這種試題可以是嚴格的邏輯推論，也可以是一般的抽象和概括。此類試題的提問方式一般是：

1.「從上文可推出以下哪個結論？」

2.「如果上述斷定是真的，以下哪項也一定是真的？」

3.「如果上述斷定是真的，那麼除了以下哪項，其餘的斷定也必定是真的？」

4.「以下哪項，作為結論從上述題幹中推出最為恰當？」

5.「下述哪項最能概括上文的主要觀點？」

【例題】

先天的遺傳因素和後天的環境影響對人的發展所起作用到底哪一個重要？雙胞胎的研究對於回答這一問題有重要的作用。惟環境影響決定論預言，如果把一對雙胞胎兒完全分開撫養，同時把一對不相關的嬰兒放在一起撫養，那麼，待他們長大成人

後，在性格等內在特徵上，前兩者之間決不會比後兩者之間有更多的類似。實際的統計數據並不支持這種極端的觀點，但也不支持另一種極端觀點，即惟遺傳因素決定論。

從以上論述最能推出以下哪個結論？

A. 為了確定上述兩種極端觀點哪一個正確，還需要進一步的研究工作。

B. 雖然不能說環境影響對於人的發展起惟一決定作用，但實際上起最重要的作用。

C. 環境影響和遺傳因素對人的發展都起著重要的作用。

D. 試圖通過改變一個人的環境來改變一個人，是徒勞無益的。

E. 雙胞胎研究是不能令人滿意的，因為它得出了自相矛盾的結論。

【答案及解釋】

答案：C

這題屬於「推出結論型」中的抽象概括結論類型。題幹所討論的問題是「先天的遺傳因素和後天的環境影響對人的發展所起作用到底哪一個重要」，在這個問題上存在著「惟環境影響決定論」和「惟遺傳因素決定論」的兩種極端的觀點，但是題幹中認為「實際的統計數據並不支持這種極端的觀點，但也不支持另一種極端觀點」，即不能片面地強調兩種因素中的任何一種，而是兩種因素都起著重要的作用。這正是 C 所表明的結論。

試題具有迷惑性的方面是，題幹用了大量的文字來敘述「惟環境影響決定論」者的預言，似乎選項 B 是正確結論，但實際上並非如此。

選項 A 不成立，因為題幹中說已經進行了實驗，而且得出了統計數據。

選項 D 與題幹不相關。因此正確答案是 C。

評價一個論證

【定義】

「評價一個論證」是指通常題幹中會進行了一個完整的論證，要求考生對其論證的可靠性、正確性、恰當性等進行評價，或者對題幹中的論證方法和方式、論證意圖和目的等進行說明。該題型的提問方式一般是：

1.「以下哪項如果為真，最能對題幹論證的有效性進行評價？」
2.「以下哪項是對上述論證方法的最為恰當的概括？」

【例題】

在經歷了全球範圍的股市暴跌的衝擊以後，T國政府宣稱，它所經歷的這場股市暴跌的衝擊，是由於最近國內一些企業過快的非國有化造成的。

以下哪項，如果事實上是可操作的，最有利於評價T國政府的上述宣稱？

A. 在宏觀和微觀兩個層面上，對T國一些企業最近的非國有化進程的正面影響和負面影響進行對比。

B. 把T國受這場股市暴跌的衝擊程度，和那些經濟情況和T國類似，但最近沒有實行企業非國有化的國家所受到的衝擊程度進行對比。

C. 把T國受這場股市暴跌的衝擊程度，和那些經濟情況和T國有很大差異，但最近同樣實行了企業非國有化的國家所受到的衝擊程度進行對比。

D. 計算出在這場股市風波中T國的個體企業的平均虧損值。

E. 運用經濟計量方法預測T國的下一次股市風波的時間。

【答案及解釋】

答案：B

題幹中Ｔ國政府把「它所經歷的這場股市暴跌的衝擊」的原因歸結為「最近國內一些企業過快的非國有化」，該結論可以通過將Ｔ國「受這場股市暴跌的衝擊程度」與「那些經濟情況與Ｔ國類似，但最近沒有實行企業非國有化的國家所受到的衝擊程度」進行對比而得到。如果那些經濟情況與Ｔ國類似但最近沒有實行企業非國有化的國家，並沒有受到這場股市暴跌的衝擊或者所受到的衝擊程度較小，就可以說明Ｔ國政府的宣稱是成立的了。選項Ｃ說要將Ｔ國受這場股市暴跌的衝擊程度和那些「經濟情況與Ｔ國有很大差異並且最近同樣實行了企業非國有化的國家所受到的衝擊程度」進行對比，這只可能說明經濟情況的差異是造成受這場股市暴跌衝擊的原因。選項Ａ、Ｄ、Ｅ都不能說明題幹中Ｔ國政府的宣稱。所以，正確答案是Ｂ。

排除法

【定義】

「排除法」是通過排除與題幹一致的選項從而找到不一致的選項，或者排除不一致的選項從而找到與題幹一致的選項，進而求解答案的方法。能夠直接運用該方法的一般提問方式是：

1.「以下除哪項外，基本上表述了上述題幹的觀點？」

2.「以下哪項最可能是題幹斷定的一個反例？」

3.「以下哪項最接近於題幹斷定的含義？」

排除法在本質上就是要通過排除題幹中已經涉及的選項，進而找到題幹中未涉及的選項作為答案，或者通過排除題幹中沒有涉及的選項進而找到與題幹一致的選項作為答案，實際上在解答每一道邏輯試題時都可以試著運用排除法。

【例題】

美國政府決策者面臨的一個頭痛的問題就是所謂的「別在我家門口」綜合症。例如，盡管民意測驗一次又一次地顯示公眾大多數都贊成建新的監獄，但是，當決策者正式宣布計劃要在某地新建一所監獄時，總遭到附近居民的抗議，並且抗議者總有辦法使計劃擱淺。

以下哪項也屬於上面所說的「別在我家門口」綜合症？

A. 某家長主張，感染了愛滋病毒素的孩子不能被允許入公共學校，當知道一個感染了愛滋病毒素的孩子進入了他孩子的學校時，他立即辦理了自己孩子的退學手續。

B. 某政客主張所有政府官員必須履行個人財產公開登記，他自己遞交了一份虛假的財產登記表。

C. 某教授主張宗教團體有義務從事慈善事業，但他自己拒絕捐款資助索馬里饑民。

D. 某汽車商主張和外國進行汽車自由貿易，以有利於本國經濟，但要求本國政府限制外國制造的汽車進口。

E. 某軍事戰略家認為核戰爭會毀滅人類，但主張本國保持足夠的核能力以抵禦外部可能的核襲擊。

【答案及解釋】

答案：A

題幹中「別在我家門口」綜合症的表現形式是一種「前後言行不一致」，或者是自相矛盾：在總體上，贊成建新的監獄，另一方面，當關係到自己時，卻又反對建新的監獄。選項A中的家長是言行一致的。

選項B中的政客還是履行了自己財產公開登記的主張。

選項C中的教授自己不是宗教團體成員。

選項D中的汽車商則符合「前後言行不一致」的表現形式：在總體上，主張自由貿易，另一方面，當關係到自己時，又反對自由貿易。

試題練習：演繹推理 Q & A

請根據以下 40 條短文的內容，於每條中選出一個或一組推論。請假定短文的內容都是正確的。

1. 在一項實驗中，實驗對象的一半被編為「實驗組」，實驗組食用了大量的某種辣椒。而作為「對照組」的另一半實驗對象卻沒有吃這種辣椒。結果，實驗組的認知能力比對照組差得多。這一結果是由於這種辣椒的一種主要成分——維生素 E 造成的。

以下哪項如果為真，則最有助於證明這種辣椒中成分造成這一實驗結論？

A. 上述結論中所提到的維生素E在所有蔬菜中都有，為了保證營養必須攝入一定量這種維生素。

B. 實驗組中人們所食用的辣椒數量，是在政府食品條例規定的安全用量之內。

C. 第二次實驗時，只給一組食用大量辣椒作為實驗組，而不高於不食用辣椒的對照組。

D. 實驗前兩組實驗對象是按認知能力均等劃分的。

2. 由於近期的乾旱和高溫，導致海灣鹽度增加，引起了許多魚隻的死亡。蝦雖然可以適應高鹽度，但鹽度高也給養蝦場帶來了不幸。

以下哪項如果為真，能夠提供解釋以上現象的原因？

A. 持續的乾旱會使海灣的水位下降，這已經引起了有關機構的注意。

B. 幼蝦吃的有機物在鹽度高的環境下幾乎難以存活。

C. 水溫升高會使蝦更快速是繁殖。

D. 魚多的海灣往往蝦也多，蝦少的海灣魚也不多。

3. 為了挽救瀕臨絕種的大熊貓，一種有效的方法是把牠們都捕獲到動物園，進行人工飼養和繁殖。以下哪項如果為真，最能對上述結論提出質疑？

A. 近5年在全世界各動物園中出生的熊貓總數是9隻，而在野生自然環境中出生的熊貓的數字，不可能準確地獲得。

B. 只有在熊貓生活的自然環境中，才有牠們足夠吃的嫩竹，而嫩竹幾乎是熊貓的惟一食物。

C. 動物學家警告，對野生動物的人工飼養將會改變牠們的某些遺傳特性。

D. 提出上述觀點的是一個動物園主，他的動議帶有明顯的商業動機。

4. 在世界範圍內禁止生產各種破壞臭氧層的化學物質可能僅僅是一種幻想，因為大量這樣的化學物質已經生產出來，並且以成千上萬部雪櫃的冷卻劑的形式存在。當這些化學物質到達大氣層中的臭氧層時，其作用不可能停止。因此，沒有任何方式可以阻止這類化學物質進一步破壞臭氧層。

   以下哪項如果為真，則能最嚴重地削弱以上論證？

   A. 不可能精確地測量雪櫃裡冷卻劑這種破壞臭氧層的化學物質的量是多少。

   B. 不會破壞臭氧層的替代品還未開發出來，並且替代品可能比雪櫃目前使用的冷卻劑昂貴。

   C. 即使人們放棄使用冷藏設備，已經存在的冰箱裡的冷卻劑也是對大氣臭氧層的一個威脅。

   D. 當雪櫃的使用壽命結束時，雪櫃裡的冷卻劑可完全回收並且重新被利用。

5. 高脂肪、高糖含量的食物危害人類健康。因此，既然越來越多的國家明令禁止未成年人吸煙和喝含酒精的飲料，那麼，為什麼不能用同樣的方法對待那些有害健康的食品呢？應該明令禁止18歲以下的人食用高脂肪、高糖食品。

   以下哪項如果為真，最能削弱上述建議？

   A. 許多國家已經把未成年人的標準訂在16歲以下。

   B. 煙、酒對人體的危害比高脂肪、高糖食物的危害要大。

   C. 禁止有害健康食品的生產，要比禁止有害健康食品的食用更有效。

   D. 高脂肪、高糖食品主要危害中年人的健康。

6. 一項對某大學國際經濟系99屆畢業生的調查結果看起來有些問題，當被調查者被問及其在校時學習成績的名次時，統計資料表明，有60%的回答者說他們的成績位於班級的前20名。

   如果回答者說的都是真話，那麼，下面哪項能夠對上述現象給出更合適的解釋？

   A. 未回答者中也並不是所有人的成績都在班級的前20名之外。

   B. 雖然回答者沒有錯報成績，但不排除個別人對於學習成績的排名有不同的理解。

   C. 成績較差的畢業生在被訪問時，一般沒有回答這個有關學習成績名次的問題。

   D. 在校學習成績名次是一個敏感的問題，幾乎所有的畢業生都進行略微的美化。

7. 據最近的統計，在需要同等學歷的十個不同職業中，教師的平均工資5年前排名第九位，而目前上升到第六位；另外，目前教師的平均工資是其他上述職業的平均工資的86%，而5年前只是55%。因此，教師工資相對偏低的狀況有了較大的改善，教師的相對生活水平有了很大的提高。上述論證基於以下哪項假設？

   I. 近5年來的通貨膨脹率基本保持穩定。

   II. 和其他職業一樣，教師中的最高工資和最低工資的差別是很懸殊的。

   III. 學歷是確定工資標準的主要依據。

   IV. 工資是實際收入的主要部分。

A. I、III
B. II、IV
C. III
D. III、IV

8. 面試在求職過程中非常重要。經過面試,如果應聘者的性格不適合待聘工作的要求,則不可能被錄用。

以上論斷是建立在哪項假設的基礎上的?

A. 必須經過面試才能取得工作,這是工商界的規矩。

B. 面試主持者能夠準確地分辨出哪些性格是工作所需要的。

C. 面試的惟一目的就是測試應聘者的個性。

D. 若一個人的性格適合工作的要求,他就一定被取錄。

9. 政府應該不允許煙草公司在其營業收入中扣除廣告費用。這樣的話,煙草公司將會繳納更多的稅項。煙草公司只好提高自己的產品價格,而產品價格的提高正好可以起到減少煙草購買的作用。以下哪項是題幹論點的前提?

A. 煙草公司不可能降低其他方面的成本來抵銷多繳的稅項。

B. 如果它們需要付高額的稅項,煙草公司將不再繼續做廣告。

C. 如果煙草公司不做廣告,香煙的銷售量將受到很大影響。

D. 煙草公司由此所增加的稅項,應該等於價格上漲所增加的盈利。

10. 彭先生是一位電腦程式專家，姚小姐是一位數學家。其實，所有的電腦程式專家都是數學家。我們知道，今天多數大學都在培養電腦程式專家。據此，我們可以認為：

A. 彭先生是由大學所培養的

B. 多數電腦程式專家都是由大學培養的

C. 姚小姐並不是大學畢業生

D. 有些數學家是電腦程式專家

11. 甲、乙、丙、丁四人的血型各不相同，甲說：「我是 A 型。」乙說：「我是 O 型。」丙說：「我是 AB 型。」丁說：「我不是 AB 型。」

四個人中只有一個人的話是假的。以下哪項成立？

A. 無論誰說假話，都能推出四個人的血型情況。

B. 乙的話假，可推出四個人的血型情況。

C. 丙的話假，可推出四個人的血型情況。

D. 丁的話假，可推出四個人的血型情況。

12. 有人認為，一個國家如果能有效率地運作經濟，就一定能創造財富而變得富有；而這樣的一個國家要想保持政治穩定，它所創造的財富必須得到公正的分配；而財富的公正分配將結束經濟風險；但是，風險的存在正是經濟有效率運作的不可或缺的先決條件。

從上述觀點可以得出以下哪項結論？

A. 一個國家政治上的穩定和經濟上的富有不可能並存。

B. 一個國家政治上的穩定和經濟上的有效率運作不可能並存。

C. 一個富有國家的經濟運作一定是有效率的。

D. 一個政治上不穩定的國家，一定同時充滿了經濟風險。

13. 一次聚會上，米高遇到了湯姆、卡爾和喬治三個人，他想知道他們三人分別是做什麼的，但三人只提供了以下信息：三人中一位是律師、一位是推銷員、一位是醫生；喬治比醫生年齡大，湯姆和推銷員不同歲數，推銷員比卡爾年紀小。

根據上述信息米高可以推出的結論是：

A. 湯姆是律師，卡爾是推銷員，喬治是醫生。

B. 湯姆是推銷員，卡爾是醫生，喬治是律師。

C. 湯姆是醫生，卡爾是律師，喬治是推銷員。

D. 湯姆是醫生，卡爾是推銷員，喬治是律師。

14. 麥角城是一種可以在植物種子的表層大量滋生的菌類，特別多見於黑麥。麥角城中含有一種危害人體的有毒化學物質。黑麥是在中世紀引進歐洲的。由於黑麥可以在小麥難以生長的貧瘠和潮濕的土地上有較好的收成，因此，就成了那個時代貧窮農民的主要食品來源。

上述信息最能支持以下哪項斷定？

A. 在中世紀以前，麥角城從未在歐洲出現。

B. 在中世紀的歐洲，如果不食用黑麥，就可以避免受到麥角城所含有毒物質的危害。

C. 在中世紀的歐洲，富裕農民比貧窮農民較多地意識到麥角城所含有毒物質的危害。

D. 在中世紀的歐洲，富裕農民比貧窮農民較少地受到麥角城所含有毒物質的危害。

15. 某律師事務所內共有12名工作人員：(1)有人會使用電腦；(2)有人不會使用電腦；(3)所長不會使用電腦。這三個命題中只有一個是真的，以下哪項正確地表示了該律師事務所會使用電腦的人數？

A. 12人都會使用

B. 12人沒人會使用

C. 僅有一人會使用

D. 不能確定

16. 許多國家首腦在出任前都並未有豐富的外交經驗，但這並沒有妨礙他們做出過成功的外交決策。外交學院的教授告訴我們，豐富的外交經驗對於成功的外交決策是不可缺少的，但事實上，一個人只要有高度的政治敏感、準確的信息分析能力和果斷的個人勇氣，就能很快地學會如何做出成功的外交決策。

對於一個缺少以上三種素養的外交決策者來說，豐富的外交經驗沒有什麼價值。如果上述斷定為真，則以下哪項一定為真？

A. 外交學院的教授比出任前的國家首腦具有更多的外交經驗。

B. 具有高度的政治敏感、準確的信息分析能力和果斷的個人勇氣，是一個國家首腦做出成功的外交決策的必要條件。

C. 豐富的外交經驗，對於國家首腦做出成功的外交決策來說，既不是充分條件，也不是必要條件。

D. 豐富的外交經驗，對於要家首腦做出成功的外交決策來說，是必要條件，但不是充分條件。

E. 在其他條件相同的情況下，外交經驗越豐富，越有利於做出成功的外交決策。

17. 某銀行的夾萬被撬，巨額現金和證券失竊。探員拘捕了三名重大的嫌疑犯：施辛格、賴普頓和安傑士。通過審訊，查明了以下的事實：

(1) 夾萬是用專門用作犯案的工具撬開，使用這種工具必須受過專門的訓練。

(2) 如果施辛格犯案，那麼安傑士犯案

(3) 賴普頓沒有受過使用犯案工具的專門訓練。

(4) 罪犯就是這三個人的一個或一伙。

以下的結論，哪個是正確的？

A. 施辛格是罪犯，賴普頓和安傑士情況不明。
B. 施辛格和賴普頓是罪犯，安傑士情況不明。
C. 安傑士是罪犯，施辛格和賴普頓情況不明。
D. 賴普頓是罪犯，施辛格和安傑士情況不明。
E. 施辛格、賴普頓和安傑士都是罪犯。

18. 雖然防滑剎車系統確實具有某些獨特的安全性能，但統計數據顯示：有防滑剎車系統的汽車的事故發生率，反而比沒有這種系統的汽車要高。

以下各項如果為真，都能對題幹陳述的現象做出解釋，除了：

A. 大多數有防滑剎車系統汽車的司機，比普通汽車的司機在駕駛時思想更為麻痹。
B. 防滑剎車系統比普通剎車系統更易出現故障。

C. 防滑剎車系統的安全性能只在時速80公里內有效,而最嚴重的交通事故都有發生在高速公路上。

D. 大多數有防滑剎車系統汽車的司機都缺乏正確使用該系統的必要培訓。

E. 防滑剎車系統具有普通剎車系統不具有的某些特殊安全性能,但同時需要昂貴的特殊維修,才能達到普通剎車系統一般維修就能達到的水平。

19. 在一些沼澤地區中,帶劇毒的鏈蛇和一些無毒蛇一樣,在蛇皮表面都有紅白黑相間的鮮艷花紋。而就在離沼澤地不遠的乾燥地帶,鏈蛇的花紋中沒有了紅色;奇怪的是,這些地區的無毒蛇的花紋中同樣沒有了紅色。

對這種現象的一個解釋是:在上述沼澤和乾燥地帶中,無毒蛇為了保護自己,在進化過程中逐步變異為具有和鏈蛇相似的體表花紋。

以下哪項最可能是上述解釋所假設的?

A. 毒蛇比無毒蛇更容易受到攻擊。

B. 在乾燥地區,紅色是自然界中的一種常見色,動物體表的紅色較不容易被發現。

C. 鏈蛇體表的顏色對其捕食的對象有很強的威懾作用。

D. 以蛇為食物的捕獵者盡量避免捕捉劇毒的鏈蛇,以免在食用時發生危險。

E. 蛇在乾燥地帶比在沼澤地帶更易受到攻擊。

20. 某珠寶店失竊，五個職員涉嫌被拘審。假設這五個職員中，
    參與犯案的人說的都是假話，無辜者說的都是真話。這五個
    職員分別有以下供述：

    張說：「王是犯案者。王說過他犯的案。」

    王說：「李是犯案者。」

    李說：「是趙犯的案。」

    趙說：「是孫犯的案。」

    孫沒有說一句話。

    依據以上的敘述，能推斷出以下哪項結論？

    A. 張犯案，王沒有犯案，李犯案，趙沒犯案，孫犯案。
    B. 張沒犯案，王犯案，李沒犯案，趙犯案，孫沒犯案。
    C. 五個職員都參與犯案。
    D. 五個職員都沒有犯案。
    E. 題幹中缺乏足夠的信息來確定每個職員是否犯案。

21. 張先生的身體狀況恐怕不宜繼續擔任部門經理的職務。因為
    近一年來，只要張先生給總經理寫信，內容只有一個：不是
    這裡不舒服，就是那裡有毛病。

    為了要使上述的論證成立，以下哪項是必須假設的？

    (1) 勝任部門經理的職務，需要良好的身體條件。

    (2) 張先生給總經理的信的內容基本上都是真實的。

    (3) 近一年來，張先生經常給總經理寫信。

A. 只有(1)

B. 只有(2)

C. 只有(3)

D. 只有(1)和(2)

E. (1)、(2)、和(3)

22. 愛爾蘭有大片泥煤蘊藏量豐富的濕地。環境保護主義者一直反對在濕地區域採煤。他們的理由是開採泥煤會破壞愛爾蘭濕地的生態平衡,其直接嚴重後果是會污染水源。這一擔心是站不住腳的。據近50年的相關統計,從未發現過因採煤而污染水源的報告。

以下哪能項如果為真,最能加強題幹的論證?

A. 在愛爾蘭的濕地採煤已有200年的歷史,其間從未因此造成水源污染。

B. 在愛爾蘭,採煤濕地的生態環境和未採煤濕地沒有實質的不同。

C. 在愛爾蘭,採煤濕地的生態環境和未開採前沒有實質性的不同。

D. 愛爾蘭具備足夠的科技水平和財政支持來治理污染,保護生態。

E. 愛爾蘭是世界上生態環境最佳的國家之一。

23. 在20世紀30年代，人們已經發現了一種有綠色和褐色纖維的棉花。但是，真到最近培育出一種可以機紡的長纖維品種後，它們才具有了商業上的價值。由於這種棉花不需要染色，加工企業就省去了染色的開銷，並且避免了由染色工藝流程帶來的環境污染。

從上述題幹可以推出下面哪項結論？

(1) 只能手紡的綠色或褐色棉花不具有商業價值。

(2) 短纖維的綠色或褐色棉花只能手紡。

(3) 在棉花加工中如果省去了染色就可以避免造成環境污染。

A. 只有(1)和(2)

B. 只有(1)和(3)

C. 只有(2)和(3)

D. 只有(1)

E. (1)、(2)和(3)

24. 一個足球教練這樣教導他的隊員：「足球比賽從來是以結果論英雄。在足球比賽中，你不是贏家就是輸家；在球迷的眼裡，你要麼是勇敢者，要麼是懦弱者。由於所有的贏家在球迷眼裡都是勇敢者，所以每個輸家在球迷眼裡都是懦弱者。」

為使上述足球教練的論證成立，以下哪項是必須假設的？

A. 在球迷看來，球場上勇敢者必勝。

B. 球迷具有區分勇敢和懦弱的準確判斷力。

C. 球迷眼中的勇敢者，不一定是真正的勇敢者。

D. 即使在球場上，輸贏也不是區別勇敢和懦弱的唯一標準。

E. 在足球比賽中，贏家一定是勇敢者。

25. 在漢語和英語中，「塔」的發音是一樣的，這是英語借用了漢語；「幽默」的發音也是一樣的，這是漢語借用了英語。而在英語和姆巴拉拉語中，「狗」的發音也是一樣的，但可以肯定，使用這兩種語言的人的交往只是近兩個世紀的事，而姆巴拉拉語(包括「狗」的發音)的歷史，幾乎和英語一樣古老。

另外，這兩種語言，屬於完全不同的語系，沒有任何親緣關係。因此，這說明，不同的語言中出現意義和發音相同的詞，並不一定是由於語言間的相互借用，或是由於語言的親緣關係所致。

上述論證必須假設以下哪項？

A. 漢語和英語中，意義和發音相同的詞都是互相借用的結果。

B. 除了英語和姆巴拉拉語以外，還有多種語言對「狗」有相同的發音。

C. 沒有第三種語言從英語或姆巴拉拉語中借用「狗」一詞。

D. 如果兩種不同語系的語言中有的詞發音相同，則使用這兩種語言的人一定在某個時期彼此接觸過。

E. 使用不同語言的人互相接觸，一定會導致語言的互相借用。

26. 西雙版納植物園中有兩種櫻草：一種是自花授粉，另一種是非自花授粉，即須依靠昆蟲授粉。近幾年來，授粉昆蟲的數量顯著減少。另外，一株非自花授粉的櫻草所結的種子比自花授粉的要少。顯然，非自花授粉櫻草的繁殖條件比自花授粉的要差。但是，遊人在植物園多見的是非自花授粉櫻草而不是自花授粉櫻草。以下哪項斷定最無助於解釋上述現象？

A. 和自花授粉櫻草相比，非自花授粉櫻草的種子發芽率較高。

B. 非自花授粉櫻草是本地植物，而自花授粉櫻草是前幾年從國外引進的。

C. 前幾年，上述植物園中非自花授粉櫻草和自花授粉櫻草的數量比大約是5比1。

D. 當兩種櫻草雜生時，土壤中的養分更易於被非自花授粉櫻草吸收，這又往往導致自花授粉櫻草的枯萎。

E. 在上述植物園中，為保護授粉昆蟲免受游客傷害，非自花授粉櫻草多植於園林深處。

27. 上個世紀60年代初以來，新加坡的人均預期壽命不斷上升，到本世紀已超過日本，成為世界之最。與此同時，和一切發達國家一樣，由於飲食中的高脂肪含量，新加坡人的心血管疾病的發病率也逐年上升。從上述斷定，最可能推出以下哪項結論？

A. 新加坡人的心血管疾病的發病率雖然逐年上升，但這種疾病不是造成目前新加坡人死亡的主要殺手。

B. 目前新加坡對於心血管病的治療水平是全世界最高的。

C. 上個世紀60年代造成新加坡人死亡的那些主要疾病，到本世紀，如果在該國的發病率沒有實質性的降低，那麼對這些疾病的醫治水平一定有實質性的提高。

D. 目前新加坡人心血管疾病的發病率低於日本。

E. 新加坡人比日本人更喜歡吃脂肪含量高的食物。

28. 有人養了一些兔子。別人問他有多少隻雌兔，多少隻雄兔？他答：在他所養的兔子中，每一隻雄兔的雌性同伴比牠的雄性同伴少一隻；而每一隻雌兔的雄性同伴比牠的雌性同伴的兩倍少兩隻。

根據上述回答，可以判斷他養了多少隻雌兔？多少隻雄兔？

A. 8隻雄兔，6隻雌兔。

B. 10隻雄兔，8隻雌兔。

C. 12隻雄兔，10隻雌兔。

D. 14隻雄兔，8隻雌兔。

E. 14隻雄兔，12隻雌兔。

29. 某出版社近年來出版物的錯字率較前幾年有明顯的增加，引起讀者的不滿和有關部門的批評，這主要是由於該出版社大量引進非專業編輯所致。當然，近年來該社出版物的大量增加也是一個重要原因。

上述議論中的漏洞，也類似地出現在以下哪項中？

Ⅰ. 美國航空公司近兩年來的投訴率比前幾年有明顯的下降。這主要是由於該航空公司在裁員整頓的基礎上，有效地提高了服務質量。當然，911事件後航班乘客數量的銳減也是一個重要原因。

Ⅱ. 統計數字表明：近年來香港心血管病的死亡率，即由心血管病導致的死亡在整個死亡人數中的比例，較前有明顯增加，這主要是由於隨著經濟的發展，香港人的飲食結構和生活方式發生了容易誘發心血管病的不良變化。當然，由於心血管病主要是老年病，因此，香港人口的老齡化，即人口中老年人比例的增大也是一個重要原因。

Ⅲ. S市今年的高考錄取率比去年增加了15%，這主要是由於各中學嚴格監控了教育質量。當然，另一個重要原因是，該市今年參加高考的人數比去年增加了20%.

A. 只有Ⅰ

B. 只有Ⅱ

C. 只有Ⅲ

D. 只有Ⅰ和Ⅲ

E. Ⅰ、Ⅱ和Ⅲ

30.宏達汽車公司生產的小轎車都安裝了駕駛員安全氣囊。在安裝駕駛員安全氣囊的小轎車中，有50%安裝了乘客安全氣囊。只有安裝乘客安全氣囊的小轎車才會同時安裝減輕衝擊力的安全杠和防碎玻璃。

如果上述斷定為真，並且事實上李先生從宏達公司購進一輛小轎車中裝有防碎玻璃，則以下哪項斷定一定為真？

Ⅰ.這輛車一定裝有安全杠

Ⅱ.這輛車一定裝有乘客安全氣囊

Ⅲ.這輛車一定裝有駕駛員安全氣囊

A. 只有Ⅰ

B. 只有Ⅱ

C. 只有Ⅲ

D. 只有Ⅰ和Ⅱ

E. Ⅰ、Ⅱ和Ⅲ

31. 圖示方法是幾何學課程的一種常用方法。這種方法使得這門課比較容易學，因為學生們得到了對幾何概念的直觀理解，這有助於培養他們處理抽象運算符號的能力。對代數概念進行圖解相信會有同樣的教學效果，雖然對數學的深刻理解從本質上說是抽象的而非想像的。

上述議論最不可能支持以下哪項斷定？

A. 通過圖示獲得直觀，並不是數學理解的最後步驟。

B. 具有很強的處理抽象運算符號能力的人，不一定具有抽象的數學理解能力。

C. 幾何學課程中的圖示方法是一種有效的教學方法。

D. 培養處理抽象運算符號的能力是幾何學課程的目標之一。

E. 存在著一種教學方法，可以有效地用於幾何學，又用於代數。

32. 因為青少年缺乏基本的駕駛技巧，特別是缺乏緊急情況的應變能力，所以必須給青少年的駕駛執照附加限制。在這點上，應當吸取H國的教訓。在H國，法律規定16歲以上就可申請駕駛執照。盡管在該國註冊的司機中19歲以下的只佔7%，但他們卻是20%的造成死亡的交通事故的肇事者。

以下各項有關H國的斷定如果為真，都能削弱上述議論，除了：

A. 與其他人相比，青少年開的車較舊，性能也較差。

B. 青少年開車時載客的人數比其他司機要多。

C. 青少年開車的年均公里(即每年平均行駛的公里數)要高於其他司機。

D. 和其他司機相比，青少年較不習慣綁上安全帶。

E. 據統計，被查出酒後開車的司機中，青少年所佔的比例，遠高於他們佔整個司機總數的比例。

33. 最近台灣航空公司客機墜落事故急劇增加的主要原因是飛行員缺乏經驗。台灣航空部門必須採取措施淘汰不合格的飛行員，聘用有經驗的飛行員。毫無疑問，這樣的飛行員是存在的。但問題在於，確定和評估飛行員的經驗是不可能的。例如，一個在氣候良好的澳洲飛行 1,000 小時的教官，和一個在充滿暴風雪的加拿大東北部飛行 1,000 小時的夜班貨機飛行員是無法相比的。

上述議論最能推出以下哪項結論(假設台灣航空公司繼續維持原有的經營規模)？

A. 台灣航空公司客機墜落事故急劇增加的現象是不可改變的。

B. 台灣航空公司應當聘用加拿大飛行員，而不宜聘用澳洲飛行員。

C. 台灣航空公司應當解聘所有現職飛行員。

D. 飛行時間不應成為評估飛行員經驗的標準。

E. 對台灣航空公司來說，沒有一項措施，能根本扭轉台灣航空公司客機墜落事故急劇增加的趨勢。

34. 一個人從飲食中攝入的膽固醇和脂肪越多，他的血清膽固醇指標就越高。存在著一個界限，在這個界限內，兩者成正比。超過了這個界限，即使攝入的膽固醇和脂肪急劇增加，血清膽固醇指標也只會緩慢地有所提高。這個界限，對於各個人種是一樣的，大約是歐洲人均膽固醇和脂肪攝入量的1/4。

上述斷定最能支持以下哪項結論？

A. 中國的人均膽固醇和脂肪攝入量是歐洲的1/2，但中國的人均血清膽固醇指標不一定等於歐洲人的1/2。

B. 上述界限可以通過減少膽固醇和脂肪的攝入量得到降低。

C. 3/4的歐洲人的血清膽固醇含量超出正常指標。

D. 如果把膽固醇和脂肪的攝入量控制在上述界限內，就能確保血清膽固醇指標的正常。

E. 血清膽固醇的含量只受飲食的影響，不受其他因素，例如運動、吸煙等生活方式的影響。

35. S市餐飲業經營點的數量自1996年的約20,000個，逐年下降至2001年的約5,000個。但是，這5年來，該市餐飲業的經營資本在整個服務行業中所佔的比例並沒有減少。

以下各項中，哪項最無助於說明上述現象？

A. S市2001年餐飲業的經營資本總額比1996年略高。

B. S市2001年餐飲業經營點的平均經營資本額比1996年有顯著增長。

C. 作為激烈競爭的結果，近5年來，S市的餐館有的被迫停業，有的則努力擴大經營規模。

D. 1996年以來，S市服務行業的經營資本總額逐年下降。

E. 1996年以來，S市服務行業的經營資本佔全市產業經營總資本的比例逐年下降。

36. 建築歷史學家丹尼斯教授對歐洲19世紀早期鋪有木地板的房子進行了研究，結果發現較大的房間鋪設的木板條比較小房間的木板條窄得多。丹尼斯教授認為，既然大房子的主人一般都比小房子的主人富有，那麼，用窄木條鋪地板很可能是當時有地位的象徵，用以表明房主的富有。

以下哪項如果為真，最能加強丹尼斯教授的觀點？

A. 歐洲19世紀晚期的大多數房子所鋪設的木地板的寬度大致相同。

B. 丹尼斯教授的學術地位得到了國際建築歷史學界的公認。

C. 歐洲19世紀早期，木地板條的價格是以長度為標準計算的。

D. 歐洲19世紀早期，有些大房子鋪設的是比木地板昂貴得多的大理石。

E. 在以歐洲19世紀市民生活為背景的小說《霧都十三夜》中，富商查理的別墅中鋪設的就是有別於民間的細條胡桃木地板。

37. 不必然任何經濟發展都會導致生態惡化，但不可能有不阻礙經濟發展的生態惡化。以下哪項最為準確地表達了題幹的含義？

A. 任何經濟發展都不必然導致生態惡化，但任何生態惡化都必然阻礙經濟發展。

B. 有的經濟發展可能導致生態惡化，而任何生態惡化都可能阻礙經濟發展。

C. 有的經濟發展可能不導致生態惡化，但任何生態惡化都可能阻礙經濟發展。

D. 有的經濟發展可能不導致生態惡化，但任何生態惡化都必然阻礙經濟發展。

E. 任何經濟發展都可能不導致生態惡化，但有的生態惡化必然阻礙經濟發展。

38. 一個美國議員提出必須對本州不斷上升的監獄費用採取措施，他的理由是：現時一個關在單人牢房裡的犯人，平均每日所需的費用就高達132美元，即使在世界上開銷最昂貴的城市裡，也不難在最好的酒店裡找到每晚租金低於125美元的房間。

以下哪項如果為真，能構成對上述美國議員的觀點及其論證的恰當駁斥？

I. 據州司法部公佈的數字，一個關在單人牢房裡的犯人所需的費用，平均每天125美元。

II. 在世界上開銷最昂貴的城市裡，很難在最好的酒店裡找到每晚租金低於125美元的房間。

III. 監獄用於犯人的費用和酒店用於客人的費用，幾乎用於

完全不同的開支項目。

A. 只有I

B. 只有II

C. 只有III

D. 只有I和II

E. I、II和III

39. 天文學家一直假設，宇宙中的一些物質是看不見的。研究顯示：許多星雲如果都是由能看見的星球構成的話，它們的移動速度要比在任何條件下能觀測到的快得多。專家們由此推測：這樣的星雲中包含著看不見的巨大質量的物質，其重力影響著星雲的運動。

以下哪項是題幹的議論所假設的？

I. 題幹所說的看不見，是指不可能被看見，而不是指離地球太遠，不能被人的肉眼或借助天文望遠鏡看見。

II. 上述星雲中能被看見的星球的總體質量可以得到較為準確的估計。

III. 宇宙中看不見的物質，除了不能被看見這點以外，具有看得見的物質的所有屬性，例如具有重力。

A. 只有I

B. 只有II

C. 只有III

D. 只有I和II

E. I、II和III

40. 家用電爐有三個部件：加熱器、恆溫器和安全器。加熱器只有兩個設置：開和關。在正常工作的情況下，如果將加熱器設置為開，則電爐運作加熱功能；設置為關，則停止這一功能。當溫度達到恆溫器的溫度旋鈕所設定的讀數時，加熱器自動關閉。電爐中只有恆溫器具有這一功能。只要溫度一超出溫度旋鈕的最高讀數，安全器自動關閉加熱器。同樣，電爐中只有安全器具有這一功能。當電爐啟動時，三個部件同時工作，除非發生故障。

以上斷定最能支持以下哪項結論？

A. 一個電爐，如果它的恆溫器和安全器都出現了故障，則它的溫度一定會超出溫度旋鈕的最高讀數。

B. 一個電爐，如果其加熱的溫度超出了溫度旋鈕的設定讀數但加熱器並沒有關閉，則安全器出現了故障。

C. 一個電爐，如果加熱器自動關閉，則恆溫器一定工作正常。

D. 一個電爐，如果其加熱的溫度超出了溫度旋鈕的最高讀數，則它的恆溫器和安全器一定都出現了故障。

E. 一個電爐，如果其加熱的溫度超出了溫度旋鈕的最高讀數，則它的恆溫器和安全器不一定都出現了故障，但至少其中某一個出現了故障。

答案與解釋

1. D

本題屬於前提型題目。讓考生找出支持實驗結論的選項，我們看看題幹中結論是實驗組認知能力不如對照組，小前提是實驗組吃了含維生素E的辣椒，而對比組沒有吃。找出大前提，我們從選項可以知道，D選項是說兩組的認知能力實驗前均等劃分了，這樣結論就會是因為吃辣椒所引起的，所以，只有D是有助於證明實驗的結論。其他選項都無助於證明實驗結論，所以選擇D。

2. B

本題是解釋型題目。題幹給出了「蝦雖然適應高鹽度，但卻給養蝦場帶來了不幸」似乎矛盾的現象，具體是什麼原因造成的呢？肯定還存在一個原因使蝦在這樣的環境蝦難以生存，我們看選項：A是無關選項，排除掉；B是可以說明這個結論，應該是正確答案；C不僅不能解釋，反而更矛盾了，所以排除掉；D也無關，所以排除掉；因此，我們選擇B。

3. B

本題是削弱型題目。題幹說為了挽救大熊貓可以把他們在動物園中進行人工養殖，要想質疑這種做法，就是要找到動物園中的環境不能使熊貓適應，難以生存。我們看選項：A是不能置疑的，排除掉；C是一個不確定的論斷，所以不能質疑論述，排除掉；D聽起來像能質疑，但是卻不是問題的關鍵所在，只有B才是最可能的原因，因為缺少自然環境中的惟一食物嫩竹，熊貓是無法生存的，所以選擇B。

4. D

本題是屬於削弱型題目。我們直接找與題幹唱反調的選項。題幹主要是說沒有辦法阻止大量以雪櫃冷卻劑形式存在化學物質破壞大氣臭氧層。而如果我們採取措施使雪櫃中的冷卻劑不進入大氣層，那麼就不能破壞臭氧層了，就削弱了題幹，而選項D正好是解決這個問題的方法，其所以我們可以直接選項D。

5. D

本題屬於削弱型題目。題幹建議明令禁止18歲以下的人食用危害人們健康的高脂肪、高糖食品。要想削弱這個建議，就必須說明這些高脂肪、高糖食品對18歲以下的人危害不大，可見只有D是說這些食品主要危害中年人健康，對18歲以下的少年危害不大，所以沒有必要明令禁止食用，所以，選擇D。

6. D

本題是解釋型題目。題幹中給出了「當被調查者被問及其在校時學習成績的名次時，有60%的回答者說他們的成績位於班級的前20名」似乎矛盾的現象，是什麼原因造成這種現象呢？選項A不能解釋這種現象，反而使之更矛盾了；B與題幹無關，排除掉；選項C也不能解釋，排除掉；只有選項D能夠解釋這個矛盾現象，因為大家都對成績敏感，回答時會美化，所以有可能不是前20名的學生也說是前20名，所以選擇D。

7. D

本題是前提型題目；題幹中的結論是「教師工資相對偏低的狀況有了較大的改善，教師的相對生活水平有了很大的提高」，從題幹中教師與其他同等學歷的九個職業工資對比情況看，我們應該假設學歷是確定工資標準的主要依據，而教師生活水平有了較大提高我們應假設工資是實際收入的主要部分，只有這樣才能得出這個結論，否則就談不上教師工資偏低和生活水平得到提高。所以正確答案是D。

8. B

本題屬於假設前提型題目。題幹中從「應聘者的性格不適合待聘工作的要求」推出「不可能被錄用」得結論，還需要假設一個大前提「面試主持者能夠準確地分辨出哪些個性是工作所需要的」，即要找到一個能夠把題幹中得小前提和結論聯結起來得選項。

選項A是無關項，C是講面試的目的，D是說的意思和題幹的前提一樣，都不能把前提和結論聯結起來，不能起到保證題幹論證能夠成立的作用。所以正確答案是B。

9. A

本題是前提型題目，使用了「三段論」，論證的結論是政府應該不允許煙草公司在營業收入中扣除廣告費用。整理成「政府不允許煙草公司扣廣告費」。前提是煙草公司會提高產品價格抵銷多繳稅項。我們要找另一個大前提，很顯然應該是煙草公司不可能通過其他渠道來抵銷多繳稅項，所以A是我們找的前提條件。B、C、D選項均為干擾項，不是前提，排除掉。所以選擇A。

10. D

只有結論D是由陳述句「所有的電腦程式專家都是數學家」直接推出來的，並不需要附帶任何假設和補充而得出的結論，因此，D是正確答案。

11. B

應用圖表代入法解題。

血型	人	
A	甲	是
B	乙	是
O	丙	是
AB	丁	非

把選項代入逐個驗證。選項B說乙說了假話，說明乙不是O型，其他三人全是真話，那麼甲說真話是A型；丙說真話是AB型；丁說真話不是AB型，那麼可能乙和丁分別是O型和B型，這樣不矛盾；再看C，如果丙說了假話，那麼丙就不是AB型，其他都是真話，那麼甲是A型，乙是O型，丁不是AB型，因為四個人血型不相同，所以丙和丁都不是AB型，是不可能的，所以有矛盾，排除掉；選項D，丁說假話，那麼丁是AB型，其他三人是真話，丙也是AB型，所以矛盾，也要排除，可見A也應排除掉，所以只有B項成立。因此選擇B。

12. B

本題是屬於假言連鎖推理型題目。題幹中的幾個命題可以分別表達為：

(1)如果有效率運作經濟，那麼創造財富而變富有；

(2)如果想保持政治穩定，那麼財富公正分配；(財富公正分配是政治穩定的必要條件)

(3)如果財富公正分配，那麼結束經濟風險；

(4)如果經濟效率運作，那麼風險存在；(風險存在是經濟有效運作的必要條件)

從上述四個命題出發可以進行如下推論：

(5) 如果想保持政治穩定，那麼結束經濟風險；(2、3連鎖推理)

(6) 如果風險不存在，那麼經濟不能有效率運作；(充分條件命題否定後件，可得到否定前件)

(7) 如果想保持政治穩定，那麼經濟不能有效率運作。

所以政治穩定與經濟有效率運作不能同時並存。選擇B。

13. C

應從題幹出發，喬治比醫生大，説明他不是醫生，排除A；湯姆和推銷員不同歲，説明湯姆不是推銷員，排除B；推銷員比卡爾年紀小，説名卡爾也不是推銷員，排除D，所以我們看，不用再繼續推斷了，只剩下C是正確的選項。

14. D

本題是結論型的題目。題幹前提説「含有麥角城這種有害物質的黑麥可以在小麥難以生長的貧瘠和潮濕的土地上有較好的收成，因此，就成了那個時代貧窮農民的主要食品來源」，還要注意時間限制，是中世紀前引入歐洲的。所以我們看選項的結論必然應該是富裕農民比貧窮農民較少受到麥角城有害物質的毒害，因為黑麥是中世紀歐洲貧窮農民的主要食品來源，所以只有D是題幹信息最支持的選項，其他都不是題幹中推出的結論，所以選擇D。

15. A

題幹中的(1)和(2)是屬於下反對關係，互為下反對關係的命題可以同真，但是不能同假。根據題意，三個命題中

只有一個是真的，那麼(1)和(2)必然有一真一假，那麼(3)就肯定是假的，説明所長會使用電腦，那麼(1)就是真的，(2)就是假的，所以可以推斷出該所12人全會使用電腦。所以選擇A。

16. C

題幹斷定：許多國家首腦在出任前都並未有豐富的外交經驗，但這並沒有妨礙他們做出成功的外交決策。這説明，豐富的外交經驗不是成功外交決策的必要條件；題幹又斷定：對於一個缺少高度的政治敏感等三種素養的外交決策者來説，豐富的外交經驗沒有什麼價值。這説明，豐富的外交經驗不是成功外交決策的充分條件，因此，如果題幹的斷定為真，則C項一定為真。

C項成立，所以D項不成立。

題幹斷定：一個人，只要有高度的政治敏感、準確的信息分析能力和果斷的個人勇氣，就能很快地學會如何做出成功的外交決策。這是斷定：這三種素養是一個國家腦做出成功外交決策的充分條件，但沒有斷定它們同時也是必要條件，因此，B項不一定為真。

題幹斷定：對於一個缺少高度的政治敏感等三種素養的外交決策者來説，豐富的外交經驗沒有什麼價值。但是，如果具備這三種素養，豐富的外交經驗有什麼價值，題幹沒做任何斷定。因此，E項不一定為真。

A項明顯屬於案外案。

17. C

從條件(1)和條件(3)，可知賴普頓是否參與犯案屬情況不明；再結合條件(4)，可知不管賴普頓是否參與犯案，三人中其它的兩個人施辛格和安傑士至少有一人參與了罪行；從條件(2)「如果施辛格犯案，那麼安傑士犯案」，可知如果安傑士不犯案，那麼施辛格也不犯案，那就沒有人參與犯案了，這與條件(4)相矛盾，所以，安傑士必須犯案。至於施辛格是否參與犯案，從所給的條件中無法作出明確的判斷。

18. E

題幹提出了一個矛盾的現象：有防滑剎車系統的事故發生率，反而比沒有這種系統的汽車要高。選項A、B、C、D都有可以從不同角度對這一矛盾現象作出說明或解釋。唯有選項E離開與交通事故的相關因素(選項A、D說的是司機的因素，選項B、C說的是汽車自身的因素)，僅僅談及維修的價格和費用等與事故無關的問題，按題意要求，該選項成立。

19. D

由選項D的斷定可知，劇毒的鏈蛇較少受到捕獵。無毒蛇在進化的過程中，逐步變異為和鏈蛇具有相似的體表花紋，這樣，就能使捕獵者誤認為是毒蛇而不加以捕獵，從而達到保護自己的目的。因此，選項D是題幹的解釋所最可能假設的。

選項A和B有可能是題幹的解釋所假設的。因為它們所作的斷定都會削弱題幹解釋的說服力。

選項C和E雖然不會削弱題幹解釋的說服力，但不會增加這種說服力，因此，不大可能是題幹的解釋所必須假設的。

20. A

由題幹，張說：「王是犯案者。王說過他作的案。」

題幹又假設：五個職員中，參與犯案的人說的都是假話，只有無辜者才說真話。所以，王如果是犯案者，他不可能承認他作的案，否則，犯案者就會說真話了。

由此可知，張說的是假話，因而張是犯案者，王不是犯案者；

由此可知，王說的是真話，因而李是犯案者；

由此可知，李說的是假話，因而趙不是犯案者；

由此可知，趙說的真話，因而孫是犯案者；

因此，選A成立。

21. E

題幹中的論證包括兩步：第一，是依據張先生給總經理的信的內容，來論證張先生的身體狀況不好；第二，是依據張先生的身體狀況不好，來論證他不宜繼續擔任部門經理的職務。

(1)必須成立，否則題幹的第二步論證就不能成立。

(2)必須成立，否則題幹的第一論證就不能成立。

(3)也必須成立，否則即使張先生給總經理的信的內容基本上都是真的，也不能得出張先生身體狀況不好的結證，使得題幹的第二步論證不能成立。

22. C

題幹要論證的結論是：環境保護主義者關於採煤會破壞愛爾蘭濕地的生態平衡的得心是站不住腳的。在諸選項中，只有C項如果為真，能得出結論：

在濕地採煤並沒有改變生態環境。其他各項都有能加強題幹的論證，但都不能得出這一結論。例如，A項如果為真，則從過去濕地採煤沒有造成水源污染，並不能得出將來濕地採煤不會造成水源污染。

再例如，B項如果成真，並不能保證在濕地採煤不改變生態環境，因為無法排除這種可能性：採煤濕地的生態環境雖然和未採煤濕地沒有實質性的不同，但卻和自身未開採的生態環境有實質性的不同。

23. A

題幹中「直到最近培育出一種可以機紡的長纖維品種後，它們才具有了商業上的價值」表達的是必要條件關係，可以整理為「只有可以機紡的長纖維，才具有商業上的價值。」根據這個條件關係，可知(1)成立，因為對必要條件假言推理說，否定前件就可以否定後件。

(2)也成立，從「直到最近培育出一種可以機紡的長纖維品種」這句話，可知以前的短纖維的綠色或褐色棉花只能手紡，而不能機紡。

(3)不一定成立，因為在棉花加中染色可能只是造成環境污染的一個工序或環節，其他的工序環節也會在棉花加工中造成環境污染。

24. A

為使足球教練的論證成立，A項是必須假設的。否則，如果在球迷看來，贏家都是勇敢者，但勇敢者不一定都是贏家，也就是說，輸家中也可能有勇敢者，這樣就不能得出結論：每個輸家在球迷眼裡都是懦弱者，即足球教練的論證不能成立。

足球教練的結論是「每個輸家在球迷眼裡都是懦弱者」。根據足球教練的論證所依據的條件，這一結論的成立不依賴於球迷判斷力的準確性，也不依賴於贏家或輸家事實上是否為勇敢者或懦弱者，因此，其餘各項均不是必須假設的。

25. C

C項是題幹的論證所必須假設的。否則，存在第三種語言從英語或姆巴巴拉拉語中借用「狗」一詞，這樣，雖然，「狗」在英語和姆巴巴拉語中的同音同義不是這兩種語言間的直接借用，但卻是通過第三種語言的間接借用。這樣，題幹的論證就難以成立。

26. E

如果E項斷定為真，則由於非自花授粉櫻草多植於園林深處，較不易被遊人看見，因此，無助於解釋為什麼遊人在植物園多見的是非自花授粉櫻草而不是自花授粉櫻草。其餘各項都從不同角度有助於對此作出解釋。例如，A項和D項斷定非自花授粉櫻草比自花授粉櫻草有更強的生命力。

27. C

假設C項的斷定不成立，即假設上個世紀60年代造成新加坡人死亡的那些主要疾病，到本世紀，在該國的發病率沒有實質性的降低，並且對這些疾病的醫治水平也沒有實質性的提高，那麼，新加坡的人均預期壽命不可能不斷上升，更難以在本世紀初成為世界之最。這說明，如果題幹的斷定為真，則C項為真，即從題幹可以推出C項。

其餘各項均不能從題幹推出。例如，A項不能從題幹推出。因為盡管新加坡的人均預期壽命是世界之最，但心血管病仍完全可能是造成目前新加坡人死亡的主要殺手。

28. A

假設雄兔的數量為 x，雌兔的數量為 y，則由條件，每一隻雄兔的雌性同伴比牠的雄性同伴少一隻，即：

(x - 1)- y = 1 (1)

每一隻雌兔的雄性同伴比牠的雌性同伴的兩倍少兩隻，即：

2(y - 1)- x = 2 (2)

由(1)式和(2)式，就可以計算得到：x = 8；y = 6。

29. D

錯字率是單位數量的文字中出現錯字的比例，一般地説，它和文字的總量沒有確定關係。題幹把近年來上述出版社出版物的大量增加，解釋為該社近年來出版物的錯字率明顯增加的重要原因，是一個漏洞。

類似地，航空公司的投訴率，是單位數量航班乘客中投訴者的比例，一般地説，它和乘客的總量沒有確定關係。選項Ⅰ把911事件後航班乘客數量的鋭減，解釋為美國航空公司投訴率有明顯下降的重要原因，是一個類似於題幹的漏洞；顯然，類似的漏洞也出現在選項Ⅲ的議論中。選項Ⅱ的議論是成立的，其中不出現類似於題幹的漏洞。

30. C

由題幹，宏達汽車公司生產的小轎車都安裝了駕駛員安全氣囊；又李先生的車是從宏達公司購進的。因此，這輛車一定裝有駕駛員安全氣囊。即選項Ⅲ為真。

選項Ⅰ和Ⅱ顯然不一定為真。

31. B

題幹斷定，對代數概念和幾何概念進行圖解有助於培養學生處理抽象運算符號的能力，至於這種處理抽象符號的能力，和對數學的深刻理解之間的關係，即和抽象的數學理解能力之間的關係，題幹未作斷定，既未作肯定性的斷定，也未作否定性的斷定。因此，題幹不支持B項。題幹支持其餘各項。例如，題幹斷定，對數學的深刻理解從本質上説是抽象的而非想像的。這説明，通過圖示獲得直觀，並不是數學理解的最後步驟。因此，題幹支持A項。

32. B

題幹的結論是：必須給青少年的駕駛執照附加限制。題幹的論據是：青少年缺乏基本的駕駛技巧，特別是缺乏緊急情況的應變能力。題幹以H國的實例來加強其論據：在該國註冊的司機中19歲以下的只佔7%，但他們卻是20%的造成死亡的交通事故的肇事者。

A、D和E項如果為真，則説明造成青少年駕車事故的一些原因，並非是他們缺乏基本的駕駛技巧，也並非是他們缺乏緊急情況的應對能力。這就削弱了題幹的議論。

C項如果為真，則説明青少年駕車事故率較高的原因之一，是因為他們有較高的年均駕駛公里數。因為顯然年均駕駛公里數較高，則發生交通事故的可能性也較高。這也對題幹有所削弱。

B項即使為真，也不能削弱題幹，因為題幹的論據涉及的是造成死亡的交通事故，即造成死亡的交通事故中，有多大的比例是青少年駕車所致，而不是交通事故的死亡率，即交通事故造成的死亡人數中，有多大的比例是青少年駕車所致。

33. E

題幹斷定，台灣航空公司客機墜落事故急劇增加的主要原因是飛行員缺乏經驗。因此，要根本扭轉台灣航空公司客機墜落事故急劇增加的趨勢，必須採取措施，聘用有經驗的飛行員。但是，題幹又同時斷定，確定和評估飛行員的經驗是不可能的。因此，可以得出結論：對台灣航空公司來說，沒有一項措施，能根本扭轉台灣航空公司客機墜落事故急劇增加的趨勢，即E項成立。

題幹也斷定，毫無疑問，有經驗的飛行員是存在的。因此，對台灣航空公司來說，沒有一項措施，能確保聘用到有經驗的飛行員，不意味著不可能聘用到有經驗的飛行員，正如在數學中，不存在一條定理能確保哥德巴赫猜想一定可證，不意味著哥德巴赫猜想不可證。因此，從題幹得不出結論：台灣航空公司客機墜落事故急劇增加的現象是不可改變的，即A項不成立。其餘各項顯然不成立。

34. A

如果一個人攝入的膽固醇及脂肪和他的血清膽固醇指標無條件成正比，那麼，如果中國的人均膽固醇和脂肪攝入量是歐洲的1/2，則其人均血清膽固醇指標也等於歐洲人的1/2。但題幹斷定，以歐洲人均膽固醇和脂肪攝入量的1/4為界限，在該界限內，上述兩者成正比；超過這個界限，則不成正比。因此，可以得出結論：中國的人均膽固醇和脂肪攝入量是歐洲的1/2，但中國的人均血清膽固醇指標不一定等於歐洲人的1/2，即A項成立。

35. E

為什麼5年來S市餐飲業經營點的數量明顯下降，但該市餐飲業的經營資本在整個服務行業中所佔的比例並沒有減少？以下兩類數據有助於解釋這一現象：

第一，S市餐飲業盡管經營點的數量下降，但經營資本總額沒有減少；

第二，S市服務行業的經營資本總額下降。

A、B、C和D項分別從以上兩個方面斷定了有助於解釋題幹的數據或信息。

E項無助於説明題幹，因為S市服務行業的經營資本佔全市產業經營總資本的比例下降，並不意味S市服務行業經營資本總額的下降。

36. C

如果C項為真，則由於當時木地板條的價格是以長度為標準計算的，因此，鋪設相同面積的房間地面，窄木條要比寬木條昂貴，顯示出房主的富有。這就有力地加強了丹尼斯的觀點。

假設C項不成立，即如果當時木地板條的價格不是以長度為標準計算的，而是例如是以面積為標準計算的，那麼，鋪設相同面積的房間地面，窄木條並不比寬木條昂貴，這就無從顯示房主的富有，丹尼斯的觀點就難以成立。假設其餘各選項不成立，丹尼斯的觀點仍然可以成立。因此，其餘各項或者不加強丹尼斯的觀點，或者對丹尼斯的觀點有所加強，但力度不如C項。

37. D

「不必然任何經濟發展都會導致生態惡化」，等值於「有的經濟發展可能不導致生態惡化」；「不可能有不阻礙經濟發展的生態惡化」，等值於「任何生態惡化都必然阻礙經濟發展」。因此，答案是D。

38. C

題幹中議員的觀點及其論證的實質性缺陷，在於把兩個具有不同內容的數字進行不恰當的比較。

I和II實際上確認了這樣的比較是成立的，問題只在於如何使進行比較的數字更為精確，這顯然不得要領。因此，I和II並不能構成對題幹的恰當駁斥。

III指出題幹中的兩個數字具有不同的內容，這就點出了題幹的癥結，從而構成了對題幹的恰當駁斥。

39. D

題幹斷定，許多星雲如果都是由能看見的星球構成的話，它們的移動速度要比在任何條件下能觀測到的快得多。這裡強調的能看見的星球的移動，是在任何條件下能觀測到的移動，因此，題幹所說的看不見，是指不可能被看見，而不是指離地球太遠，不能被人的肉眼或借助天文望遠鏡看見，即I項是題幹的議論所假設的。

題幹，天文學家是由星雲的移動速度，計算出星雲的實際質量；由星雲的實際質量和星雲中能被看見星球的總體質量的差別，推測星雲中包含著看不見的巨大質量的物質。如果上述星雲中能被看見的星球的總體質量無法得到較為準確的估計，則就無從推測上述看不見的巨大質量的物質存在，因此，II項是題幹的議論所假設的。

題幹的議論假設宇宙中看不見的物質具有看得見的物質的某些屬性，例如具有重力，但並不假設具有除了不能被看見這點以外的所有屬性。III項的斷定過強了，不是題幹的議論必須假設的。

40. D

根據題幹的條件，一個電爐，如果其加熱的溫度超出了溫度旋鈕的最高讀數，則說明當溫度達到恆溫器的溫度旋鈕所設定的讀數時，加熱器並未自動關閉，即恆溫器出現了故障；同時也說明當溫度超出溫度旋鈕的最高讀數時，加熱器並未自動關閉，即安全器出現了故障。也就是說，一個電爐，如果其加熱的溫度超出了溫度旋鈕的最高讀數，則它的恆溫器和安全器一定都出現了故障。因此，D項作為題幹的結論成立。因為D項成立，所以E項不成立。

A項顯然不成立。例如在加熱器不工作的情況下，恆溫器和安全器即使都出現故障，電爐的溫度也不會超出溫度旋鈕的最高讀數。

B項不成立。因為一個電爐，如果其加熱的溫度超出了溫度旋鈕的設定讀數但加熱器未關閉，只能說明恆溫器出現故障，不能說明安全器出現故障。

C項不成立。因為一個電爐加熱器自動關閉，可能是恆溫器出現故障，但安全器工作正常。

雖然A、B、D項均有削弱作用，但是力度不如C項，所以選擇C。

II. Verbal Reasoning (English)

Verbal Test Reasoning 就是給出根據一篇 100 至 200 多字的短文，判斷題幹信息正確與否，主要考察應聘者的英語閱讀能力和邏輯判斷能力。

Verbal Test Reasoning 答案有三個選擇：

1. True：題幹的信息根據原文來判斷，是正確的。

2. False：題幹的信息根據原文來判斷，是錯誤的。

3. Can't tell，根據原文提供的信息，無法判斷對錯。

解題步驟(1)：定位與判斷

因為題量大，建議盡量控制在平均30秒內完成一題的速度。一般來說文章看懂後，基本可以直接判斷對錯的。但有時候往往是在「True」還是「Can't tell」，或者「False」和「Can't tell」間無法判斷。對於這種情況，建議無論如何都不要花超過45秒來考慮，應隨便在兩者中選擇一個。因為，考慮到我們的邏輯判斷往往並不完美，有時候將結果交給運氣未必不是件好事，而且不至於影響後面的題目。否則碰到最後一篇短文時，發現很簡單，但時間卻不夠，那就得不償失了。

以下是一些關於Verbal Test Reasoning的答題技巧：

STEP 1：定位。找出題目在原文中的出處。

1. 找出題目中的關鍵詞，最好先定位到原文中的一個段落。

2. 從頭到尾快速閱讀該段落，根據題目中的其它關鍵詞，在原文中找出與題目相關的一句或幾句話。

3. 仔細閱讀這一句話或幾句話，根據第二大步中的原則和規律，確定正確答案。

4. 要注意順序性，即題目的順序和原文的順序基本一致。

STEP 2：判斷。根據「True」、「False」及「Can't tell」的原則和規律，填寫正確答案。

解題步驟（2）：True 的特點

【定義1】

題目是原文的同義表達。通常用同義詞或同義結構。

【例題】

原文： Frogs are losing the ecological battle for survival, and biologists are at a loss to explain their demise. (青蛙失去了生存下來的生態競爭能力，生物學家不能解釋它們的死亡。)

題目： Biologists are unable to explain why frogs are dying. (生物學家不能解釋為什麼青蛙死亡。)

【答案及解析】

True (題目中的「are unable to」跟原文中的「are at a loss to」屬同義詞，題目中的「why frogs are dying」和原文中的「their demise」是同義詞，所以答案應為「True」。)

【定義2】

題目是根據原文中的幾句話做出推斷或歸納。不推斷不行，但有時有些同學會走入另一個極端，即自行推理或過度推理。

【例題1】

原文： Compare our admission inclusive fare and see how much you save. Cheapest is not the best and value for money is guaranteed. If you compare our bargain Daybreak fares, beware--------most of our competitors do not offer an all inclusive fare. (比較我們包含的費用會看到你省了很多錢。最便宜的不是最好的。如果你比較我們的價格，會發現絕大多數的競爭對手不提供一籃子費用。)

題目： Daybreak fares are more expensive than most of their competitors. (Daybreak的費用比絕大多數的競爭對手都昂貴。)

【答案及解析】

True (雖然文章沒有直接提到的費用比絕大多數的競爭對手都昂貴。但從原文幾句話中可以推斷出Daybreak和絕大多數的競爭對手相比，收費更高，但服務的項目要更全。與題目的意思一致，所以答案應為「True」。)

【例題2】

原文： For example, it has been demonstrated that rapid response leads to a greater likelihood of arrest only if responses are in the order of 1-2 minutes after a call is received by the police. When response times increase to 3-4 minutes ------still quite a rapid response-------the likelihood of an arrest is substantially reduced. (例如，只有反應時間在警察接到電話之後1-2分鐘，快速反應才會使抓住罪犯的可能性更大。當反應時間增加到3-4分鐘，仍然是非常快的反應，抓住罪犯的可能性就實質性的降低。)

題目： A response delay of 1-2 minutes may have substantial influence on whether or not a suspected criminal is caught. (1至2分鐘的反應延遲會對嫌疑犯是否被抓住產生實質性的影響。)

【答案及解析】

True (從原文的兩句話可以推斷出：1至2分鐘，抓住罪犯的可能性很大，3至4分鐘，可能性就實質性的降低。所以，1至2分鐘的反應延遲會對嫌疑犯是否被抓住產生實質性的影響，答案應為「True」。)

解題步驟 (3)：「False」的特點

【定義1】

題目與原文直接相反。通常用反義詞、not加同義詞及反義結構。
no longer / not any more / not / by no means …對比used to do sth. / until recently / as was once the case

【例題1】

原文： A species becomes extinct when the last individual dies.
(當最後一個個體死亡時，一個物種就滅亡了。)

題目： A species is said to be extinct when only one individual exists. (當只有一個個體存活時，一個物種就被說是滅亡了。

【答案及解析】

False (可以看出題目與原文是反義結構。原文說一個物種死光光，才叫滅絕，而題目說還有一個個體存活，就叫滅絕，題目與原文直接相反，所以答案為「False」。)

【例題2】

原文： It has been successfully used in the United States to

provide input into resource exploitation decisions and assist wildlife managers and there is now enormous potential for using population viability to assist wildlife management in Australia's forests. (在美國它已經成功地用於支持資源開發和幫助野生生命研究管理者。現在，在使用它對澳洲的森林中的野生生物管理上有巨大的潛力。)

題目： PVA has been used in Australia for many years. (PVA已經在澳大利亞使用多年了。)

【答案及解析】

False (原文說PVA在澳洲的研究中有巨大的潛力，即剛剛開始。題目說在澳洲已經使用多年，所以題目與原文是反義結構，故答案為「False」。)

【定義2】

原文是多個條件並列，題目是其中一個條件(出現must或only)
原文是兩個或多個情形(通常是兩種情形)都可以，常有both
…and、and、or及also等詞。以及various / varied / variety /
different / diversified / versatile等表示多樣性的詞彙。題目是「必須」或「只有」或是「單一」其中一個情況，常有must及only /
sole / one / single等詞。

【例題1】

原文： Booking in advance is strongly recommended as all Daybreak tours are subject to demand. Subject to availability, stand by tickets can be purchased from the driver. (提前預定是強烈建議的，因為所有的Daybreak

旅行都是由需求決定的。如果還有票的話，可直接向司機購買。)

題目：Tickets must be bought in advance from an authorized Daybreak agent. (票必須提前從一個認證的代理處購買。)

【答案及解析】

False (原文是提前預定、直接向司機購買都可以，是多個條件的並列。題目是必須提前預定，是必須其中一個情況。所以答案應為「False」。)

【例題2】

原文：Since the Winter Games began, 55 out of 56 gold medals in the men's Nordic skiing events have been won by competitors from Scandinavia or the former Soviet Union. (自從冬奧會開始，在男子越野滑雪項目中的56塊中的55塊金牌被來自北歐和前蘇聯的選手獲得。)

題目：Only Scandinavians have won gold medals in the men's winter Olympics. (只有北歐人獲得了冬運會男子越野滑雪項目中的金牌。)

【答案及解析】

False (原文是北歐人和前蘇聯的選手獲得了金牌，而且是獲得了56中的55塊，還有1塊不知道被誰獲得。題目是只有北歐人獲得了金牌。所以答案應為「False」。)

【定義3】

原文強調是一種「理論」(theory)、「感覺」(felt)、「傾向性」(trend / look at the possibilities of)、「期望或是預測」(it is predicted / expected / anticipated that) 等詞。而題目強調是一種「事實」, 常有 real / truth / fact / prove 等詞。

【例題1】

原文： But generally winter sports were felt to be too specialized. (但一般來說，冬季項目被感覺是很專門化的)

題目： The Antwerp Games proved that winter sports were too specialized. (Antwerp 運動會證明冬季項目是很專門化的。)

【答案及解析】

False (原文中有 feel，強調是「感覺」。題目中有 prove，強調是「事實」。所以答案應為「False」。)

【例題2】

原文： Another theory is that worldwide temperature increases are upsetting the breeding cycles of frogs. (另一種理論是世界範圍溫度的升高破壞了青蛙的生長循環。)

題目： It is a fact that frogs' breeding cycles are upset by worldwide increases in temperature. (一個事實是青蛙的生長循環被世界範圍溫度的升高所破壞。)

【答案及解析】

False (原文中有 theory，強調是「理論」。題目中有 fact，強調是「事實」，所以答案應為為「False」。)

【定義4】

原文和題目中使用了表示不同程度、範圍、頻率、可能性的詞。原文中常用typical、odds、many(很多)、sometimes(有時)及unlikely(不太可能)等詞。題目中常用special 、impossible、all(全部)、usually(通常)、always(總是)、及impossible(完全不可能)等詞。

【例題1】

原文： Frogs are sometimes poisonous. (青蛙有時是有毒的)

題目： Frogs are usually poisonous. (青蛙通常是有毒的)

【答案及解析】

False (原文中有sometimes，強調是「有時」。題目中有usually，強調是「通常」。所以答案應為「False」。)

【例題2】

原文： Without a qualification from a reputable school or university, it is unlikely to find a good job. (不是畢業於著名學校的人不太可能找到一個好的工作)

題目： It is impossible to get a good job without a qualification from a respected institution. (不是畢業於著名學校的人找到一個好的工作是完全不可能的。)

【答案及解析】

False (原文中有 unlikely，強調是「不太可能」。題目中有 impossible，強調是「完全不可能」。所以答案應為「False」。)

【定義5】

情況原文中包含條件狀語，題目中去掉條件成份。原文中包含條件狀語，如 if、unless 或 if not 也可能是用介詞短語表示條件狀語如 in、with、but for 或 exept for。題目中去掉了這些表示條件狀語的成份。這時，答案應為 False。

【例題】

原文： The Internet has often been criticized by the media as a hazardous tool in the hands of young computer users. (Internet 通常被媒體指責為是年輕的計算機用戶手中的危險工具。)

題目： The media has often criticized the Internet because it is dangerous. (媒體經常指責 Internet ，因為它是危險的。)

【答案及解析】

False (原文中有表示條件狀語的介詞短語 in the hands of young computer users, 題目將其去掉了。所以答案應為「False」。)

【定義6】

出現以下詞彙，題目中卻沒有說明：less obviously、less likely 或 less possible。

解題步驟(4)：「Can't tell」的特點

【定義1】

題目中的某些內容在原文中沒有提及。題目中的某些內容在原文中找不到依據。

【定義2】

題目中涉及的範圍小於原文涉及的範圍，也就是更具體。原文涉及一個較大範圍的範疇，而題目是一個具體概念。也就是說，題目中涉及的範圍比原文要小。

【例題1】

原文： Our computer club provides printer. (我們電腦學會提供打印機)

題目： Our computer club provides color printer. (我們電腦學會提供彩色打印機)

【答案及解析】

Can't tell (題目中涉及的概念「比原文中涉及的概念」要小。換句話說，電腦學會提供打印機，但是究竟是彩色還是黑白的，不知道或有可能，文章中沒有給出進一步的信息。所以答案應為「Can't tell」。

【例題2】

原文： Tourists in Cyprus come mainly from Europe. (到塞浦路斯旅遊的遊客主要來自歐洲。)

題目： Tourists in Cyprus come mainly from the UK. (到塞浦
路斯旅遊的遊客主要來英國。)

【答案及解析】

Can't tell (題目中涉及的概念「UK」比原文中涉及的概念「Europe」要小。原文只
說到塞浦路斯旅遊的遊客主要來自歐洲，有可能主要來自英國，也可能主要來自
歐洲的其他國家，文章中沒給出進一步的信息。所以答案應為「Can't tell」。

【定義3】

原文是某人的目標、目的、想法、願望、保證、發誓等，題目是
事實。原文中常用aim / goal / promise / swear / vow / pledge /
oath / resolve 等詞。題目中用實意動詞。

【例題1】

原文： He vowed he would never come back. (他發誓他將永不
回來)

題目： He never came back. (他沒再回來)

【答案及解析】

Can't tell (原文中說「他發誓將永不回來」，但實際怎麼樣，不知道。也可能他違
背了自己的誓言。所以答案應為「Can't tell」。)

【例題2】

原文： His aim was to bring together, once every four years,
athletes from all countries on the friendly fields of
amateur sport. (他的目的是把各國的運動員每4年一次
聚集到友好的業餘運動的賽場上。)

題目： Only amateur athletes are allowed to compete in the modern Olympics. (只有業餘運動員被允許在現代奧運會中競爭。)

【答案及解析】

Can't tell (原文中用「aim」表示「目的」，題目中用實意動詞表示「事實」。把各國的運動員聚集到友好的業餘運動的賽場上，這只是創建者的目的，實際情況如何，文章中沒説，所以答案應為「Can't tell」。)

【定義4】

原文中沒有比較級，題目中有比較級。

【例題】

原文： In Sydney, a vast array of ethnic and local restaurants can be found to suit all palates and pockets. (在悉尼，有各種各樣的餐館。)

題目： There is now a greater variety of restaurants to choose from in Sydney than in the past. (現在有更多種類的餐館可供選擇)

【答案及解析】

Can't tell (原文中提到了悉尼有各種各樣的餐館，但並沒有與過去相比，所以答案應為「Can't tell」。)

【定義5】

原文中是虛擬「would / even if」，但題目中卻是事實。(虛擬語氣看到當作沒有看到)

【定義6】

原文中是具體的數據事例，而題目中卻把它擴大化，規律化。

試題練習：Verbal Reasoning Q & A

In this test, each passage is followed by three statements (the questions). You have to assume what is stated in the passage is true and decide whether the statements are either:

True: The statement is already made or implied in the passage, or follows logically from the passage.

False: The statement contradicts what is said, implied by, or follows logically from the passage.

Can't tell: There is insufficient information in the passage to establish whether the statement is true or false.

Passage 1 (Question 1 to 5)

The Large Hadron Collider (LHC), located underneath the border of France and Switzerland, is one of the biggest pieces of machinery in the world. Its construction involved 9,000 magnets, and over 10,000 tons of nitrogen are used for its cooling processes. Scientists and engineers have spent £4.5 billion on building and underground track at CERN, the world's largest particle physics laboratory. This enormous scientific instrument will collect a huge amount of data, but only a small percentage of what is recorded will be useful. When proton atoms – travelling almost at the speed of light –

collide inside the LHC, theoretical physicists expect new forces and particles to be produced. It may even be possible to study black holes using this experiment.

1. Protons travel around the LHC at the speed of light.

2. The cost of the LHC's track was over £4.5 billion.

3. The LHC is the largest experiment ever conducted in the world.

4. The LHC was designed to study black holes.

5. The LHC uses over 10,000 tons of oxygen for its cooling processes.

Passage 2 (Question 6 to 10)

Large areas of land are needed for growing plants that will be distilled into biofuels. Producing biofuels from agricultural commodities has forced up the price of food. This is just one of the negative impacts that increased biofuel production has had on food security. In August, food scientist Sharon de Cruz demanded an immediate financial review of the current system of subsidies. Her argument is that there are more cost-efficient ways of supporting biofuels. For example, studies have indicated that genetically modifying crops will improve their suitability for producing biofuels.

6. Sharon de Cruz made a scientific recommendation based on environmental concerns.

7. Genetically modified crops produce biofuels more efficiently.

8. Too much land is required to produce biofuels.

9. Food security is improved by the increased use of biofuels.

10. Subsidies represent one way of supporting biofuels production.

Passage 3 (Question 11 to 15)

The frequency of MRSA being given as the cause of death on death certificates has been increasing significantly for several years. MRSA is an infection-causing bacterium that has developed a resistance to penicillin and many other antibiotics. MRSA infections represent a particular danger for hospital patients with weakened immune systems or open wounds.

Scientific trials are testing whether MRSA develops resistance after exposure to new drugs. A research breakthrough would herald a cure for the MRSA threat.

11. MRSA is resistant to all antibiotics.

12. MRSA-related deaths are now more common.

13. Further research is being conducted to study MRSA.

14. Penicillin is an effective treatment for MRSA.

15. MRSA can prove fatal.

Passage 4 (Question 16 to 19)

There are now several million cars in the UK using satellite navigation systems (satnav systems). These increasingly popular satnav systems mean that motorists no longer have to read maps while they are driving. There are two other major advantages: reduced journey time and reduced mileage (and thus fuel consumption) on unfamiliar routes. System improvements have made these devices much more accurate than earlier models, and today's designs are easier to use and have fewer distracting features. Although some safety surveys highlight the dangers of operating dashboard devices while driving, research conducted by one satnav manufacturer showed that nearly 70 per cent of drivers felt calmer and more focused on the road when using a satnav system.

16. Most drivers feel calmer when using a satnav system.

17. Controversy remains about the effects that satnav systems have on driver concentration.

18. Satellite navigation systems are useful for those people who can't read maps.

19. The passage suggests that a satnav system can make navigation more efficient.

Answer

1. False

The correct answer is FALSE because the specific section of the passage states【almost at light speed】.

2. False

The correct answer is FALSE because the passage states that the exact cost was £4.5 billion.

3. Can't tell

The correct answer is CANT'T TELL. The last sentence in the passage refers to the LHC as an experiment. The passage also states that the LHC is【one of the biggest pieces of machinery in the world】. There is no additional evidence in the passage from which to draw the conclusion that the LHC must be【the largest experiment ever conducted in the world】.

4. False

The correct answer is FALSE because the passage states that it may even be possible to study black holes using this experiment, however this was not the reason why the LHC was designed.

5. False

The correct answer is FALSE because the passage states that【over 10,000 tons of nitrogen are used for its cooling processes】.

6. False

The correct answer is FALSE because the passage states:【In August, food scientist Sharon de Cruz demanded an immediate financial review of the current system of subsidies】. Sharon is indeed a scientist but her recommendation is based on financial reasons rather than environmental concerns.

7. True

The correct answer is TRUE because the passage states【studies have indicated that genetically modifying crops will improve their suitability for producing biofuels】.

8. Can't tell

The correct answer is CAN'T TELL because the passage states【large areas of land are needed for growing plants that will be distilled into biofuels】. However there is no judgement or evidence provided within the passage that too much land is needed to produce biofuels.

9. False

The correct answer is FALSE. There is mention of food security as an issue, as follows: producing【biofuels from agricultural commodities has forced up the price of food. This is just one of the negative impacts that increased biofuel production has had on food security】. The passage considers there to be a negative impact, rather than improved food security.

10. True

The correct answer is TRUE. The relevant parts of the passage are as follows: 【in August, food scientist Sharon de Cruz demanded an immediate financial review of the current system of subsidies. Her argument is that there are cost-effiective ways of supporting biofuels】. In other words one way of supporting biofuels is by subsidy.

11. False

The correct answer is FALSE. The passage does not say that MRSA is resistant to 【all】 antibiotics, only to 【penicillin and many other antibiotics】.

12. True

The correct answer is TRUE because the passage states that 【the frequency of MRSA being given as the cause of death on death certificates has been increasing significantly for several years】.

13. True

The correct answer is TRUE because the passage states 【scientific trials are testing whether MRSA develops resistance after exposure to new drugs. A research breakthrough would herald a cure for the MRSA threat】.

14. False

The correct answer is FALSE because the passage refers to MRSA having 【developed a resistance to penicillin】.

15. True

The correct answer is TRUE because the passage refers to MRSA as a cause of death.

16. True

The correct answer is TRUE because the passage states: 【other research shows that nearly 70 per cent of drivers felt calmer and more focused on the road when using a satnav system】.

17. True

The correct answer is TRUE because the passage states that: 【although some safely surveys highlight the dangers of operating dashboard devices while driving, other research shows that nearly 70 per cent of drivers felt calmer and more focused on the road when using a satnav system】.

18. True

The correct answer is TRUE based on the following extract: 【motorists no longer have to read maps while they are driving】.

19. True

The correct answer is TRUE because the passage states: 【there are two other major advantages: reduced journey time and reduced mileage (and thus fuel consumption) on unfamiliar routes】.

III. Data Sufficiency Test

Data Sufficiency Test 不但要求考生具備基礎的數學知識和熟練的計算技巧，而且注重檢驗考生分析定量問題的能力，即根據已給出的數據，辨認哪些數據與問題有關，確定在何種情況下所給數據能滿足問題的要求，以此來檢驗考生的推理和綜合分析的能力。

Data Sufficiency Test 四個注意地方

在作答 Data Sufficiency Test 的題目時，考生應注意以下四點：

1. 不要鑽牛角尖，花過多時間找答案，你只要根據已給的數據找答案便可。

2. 即使發現(1)足以答題，也千萬不要倉促地選擇 A，相反應繼續審題，看看(2)是否也能單獨解題。

3. 考生應當熟悉某些必需的日常生活知識。例如某題提到閏年，我們就應該想到，閏年的二月份只有28天，而且應將這一數據考慮到原題中，不要因為數據(1)和(2)沒有提到它而將其忽略了。

4. 當涉及到幾何圖形時，千萬不要依賴試卷上給出的圖形而得出錯誤的假設和判斷。有時從圖形上看似乎並非全是按比例繪畫的。

對於不少考生往往在這部分的得分比較低，其中一個原因是對題型不熟悉；二是答題的速度太慢。為了提高考生的應試能力，以下為大家介紹一種較有系統的解題方法，可供大家平時練習和考試中使用：

(1)第一個條件能否單獨充分地回答題目中所提出的問題？

(2)第二個條件能否單獨充分地回答題目中所提出的問題？

(3)兩個條件加在一起能否充分地回答題目中所提出的問題？

一般應按著上面所說的(1)、(2)和(3)的順序進行解題，下面是採用系統分析方法的流程圖：

如果問題(1)的答案是肯定的，那麼可能的選擇答案是 A、C 或 E。再判斷問題(2)的答案，若肯定就選擇 C，否則選擇 A。

如果問題(1)的答案是否定的，那麼可能的選擇答案是 B、C、E 再判斷問題2的答案，若肯定就選擇 B，否則有兩種可能的答案即 C 或 E。

最後，再判斷問題2的答案，若肯定選擇C，否則選擇E。

採用這種解題方法，即使不能全部回答出上述三個問題。也可以用來排除其中不可能或錯誤的選項。例如，如果你僅知道問題(1)的答案是肯定的，那麼你就能排除掉選項B、C和E；如果你僅知道問題(3)的答案是肯定的，那麼，你就能排除掉選項E；如果你僅知道問題2的答案是否定的，那麼你就能排除選項D和B。

只要考生能熟悉地掌握上述解題方法，那麼在數據填充部分中獲得高分並非難事。

試題練習：Data Sufficiency Test Q & A

In this test, you are required to choose a combination of clues to solve a problem.

1. How many ewes (female sheep) in a flock of 50 sheep are black?

   (1) There are 10 rams (male sheep) in the flock.

   (2) Forty percent of the animals are black.

   A. statement (1) alone is sufficient, but statement (2) alone is not sufficient to answer the question

   B. statement (2) alone is sufficient, but statement (1) alone is not sufficient to answer the question

   C. both statements taken together are sufficient to answer the question, but neither statement alone is sufficient

   D. each statement alone is sufficient

   E. statements (1) and (2) together are not sufficient, and additional data is needed to answer the question

2. Is the length of a side of equilateral triangle E less than the length of a side of square F?

   (1) The perimeter of E and the perimeter of F are equal.

   (2) The ratio of the height of triangle E to the diagonal of square F is $2\sqrt{3} : 3\sqrt{2}$.

CHAPTER ONE
試題練習

CHAPTER TWO
模擬試卷

CHAPTER THREE
常見問題

A. statement (1) alone is sufficient, but statement (2) alone is not sufficient to answer the question

B. statement (2) alone is sufficient, but statement (1) alone is not sufficient to answer the question

C. both statements taken together are sufficient to answer the question, but neither statement alone is sufficient

D. each statement alone is sufficient

E. statements (1) and (2) together are not sufficient, and additional data is needed to answer the question.

3. If a and b are both positive, what percent of b is a?

(1) a = 3/11

(2) b/a = 20

A. statement (1) alone is sufficient, but statement (2) alone is not sufficient to answer the question

B. statement (2) alone is sufficient, but statement (1) alone is not sufficient to answer the question

C. both statements taken together are sufficient to answer the question, but neither statement alone is sufficient

D. each statement alone is sufficient

E. statements (1) and (2) together are not sufficient, and additional data is needed to answer the question

4. A wheel of radius 2 meters is turning at a constant speed. How many revolutions does it make in time T?

(1) T = 20 minutes.

(2) The speed at which a point on the circumference of the wheel is moving is 3 meters per minute.

A. statement (1) alone is sufficient, but statement (2) alone is not sufficient to answer the question

B. statement (2) alone is sufficient, but statement (1) alone is not sufficient to answer the question

C. both statements taken together are sufficient to answer the question, but neither statement alone is sufficient

D. each statement alone is sufficient

E. statements (1) and (2) together are not sufficient, and additional data is needed to answer the question

5. Are the integers x, y and z consecutive?

(1) The arithmetic mean (average) of x, y and z is y.

(2) y-x = z-y

A. statement (1) alone is sufficient, but statement (2) alone is not sufficient to answer the question

B. statement (2) alone is sufficient, but statement (1) alone is not sufficient to answer the question

C. both statements taken together are sufficient to answer the question, but neither statement alone is sufficient

D. each statement alone is sufficient

E. statements (1) and (2) together are not sufficient, and additional data is needed to answer the question

6. Is $x > 0$?

(1) $-2x < 0$

(2) $x^3 > 0$

A. statement (1) alone is sufficient, but statement (2) alone is not sufficient to answer the question

B. statement (2) alone is sufficient, but statement (1) alone is not sufficient to answer the question

C. both statements taken together are sufficient to answer the question, but neither statement alone is sufficient

D. each statement alone is sufficient

E. statements (1) and (2) together are not sufficient, and additional data is needed to answer the question

7. A certain straight corridor has four doors, A, B, C and D (in that order) leading off from the same side. How far apart are doors B and C?

(1) The distance between doors B and D is 10 meters.

(2) The distance between A and C is 12 meters.

A. statement (1) alone is sufficient, but statement (2) alone is not sufficient to answer the question

B. statement (2) alone is sufficient, but statement (1) alone is not sufficient to answer the question

C. both statements taken together are sufficient to answer the question, but neither statement alone is sufficient

D. each statement alone is sufficient

E. statements (1) and (2) together are not sufficient, and additional data is needed to answer the question

8. Given that x and y are real numbers, what is the value of x + y?

(1) $(x^2-y^2)/(x-y) = 7$

(2) $(x+y)^2 = 49$

A. statement (1) alone is sufficient, but statement (2) alone is not sufficient to answer the question

B. statement (2) alone is sufficient, but statement (1) alone is not sufficient to answer the question

C. both statements taken together are sufficient to answer the question, but neither statement alone is sufficient

D. each statement alone is sufficient

E. statements (1) and (2) together are not sufficient, and additional data is needed to answer the question

9. Two socks are to be picked at random from a drawer containing only black and white socks. What is the probability that both are white?

(1) The probability of the first sock being black is 1/3.

(2) There are 24 white socks in the drawer.

A. statement (1) alone is sufficient, but statement (2) alone is not sufficient to answer the question

B. statement (2) alone is sufficient, but statement (1) alone is not sufficient to answer the question

C. both statements taken together are sufficient to answer the question, but neither statement alone is sufficient

D. each statement alone is sufficient

E. statements (1) and (2) together are not sufficient, and additional data is needed to answer the question

10. A bucket was placed under a dripping tap which was dripping at a uniform rate. At what time was the bucket full?

(1) The bucket was put in place at 2pm.

(2) The bucket was half full at 6pm and three-quarters full at 8pm on the same day.

A. statement (1) alone is sufficient, but statement (2) alone is not sufficient to answer the question

B. statement (2) alone is sufficient, but statement (1) alone is not sufficient to answer the question

C. both statements taken together are sufficient to answer the question, but neither statement alone is sufficient

D. each statement alone is sufficient

E. statements (1) and (2) together are not sufficient, and additional data is needed to answer the question

11. Every pupil in a school was given one ticket for a concert. The school was charged a total of $6000 for these tickets, all of which were of equal value. What was the price of one ticket?

(1) If the price of each ticket had been one dollar less, the total cost would have been 1,200 less.

(2) If the price of each ticket had been $2 more, the total bill would have increased by 40%.

A. statement (1) alone is sufficient, but statement (2) alone is not sufficient to answer the question

B. statement (2) alone is sufficient, but statement (1) alone is not sufficient to answer the question

C. both statements taken together are sufficient to answer the question, but neither statement alone is sufficient

D. each statement alone is sufficient

E. statements (1) and (2) together are not sufficient, and additional data is needed to answer the question

12. What is the ratio of male to female officers in the police force in town T?

(1) The number of female officers is 250 less than half the number of male officers.

(2) The number of female officers is 1/7 the number of male officers.

A. statement (1) alone is sufficient, but statement (2) alone is not sufficient to answer the question

B. statement (2) alone is sufficient, but statement (1) alone is not sufficient to answer the question

C. both statements taken together are sufficient to answer the question, but neither statement alone is sufficient

D. each statement alone is sufficient

E. statements (1) and (2) together are not sufficient, and additional data is needed to answer the question

13. What is the value of n?

(1) 3n + 2m = 18

(2) n–m = 2n–(4 + m)

A. statement (1) alone is sufficient, but statement (2) alone is not sufficient to answer the question

B. statement (2) alone is sufficient, but statement (1) alone is not sufficient to answer the question

C. both statements taken together are sufficient to answer the question, but neither statement alone is sufficient

D. each statement alone is sufficient

E. statements (1) and (2) together are not sufficient, and additional data is needed to answer the question

14. How long did it take Henry to drive to work last Wednesday? (He did not stop on the way).

(1) If he had driven twice as fast he would have taken 35 minutes.

(2) His average speed was 30 miles per hour.

A. statement (1) alone is sufficient, but statement (2) alone is not sufficient to answer the question

B. statement (2) alone is sufficient, but statement (1) alone is not sufficient to answer the question

C. both statements taken together are sufficient to answer the question, but neither statement alone is sufficient

D. each statement alone is sufficient

E. statements (1) and (2) together are not sufficient, and additional data is needed to answer the question

15. What is the slope of line l which passes through the origin of a rectangular coordinate system?

(1) The line does not intersect with the line $y = x + 2$

(2) The line passes through the point $(3,3)$

A. statement (1) alone is sufficient, but statement (2) alone is not sufficient to answer the question

B. statement (2) alone is sufficient, but statement (1) alone is not sufficient to answer the question

C. both statements taken together are sufficient to answer the question, but neither statement alone is sufficient

D. each statement alone is sufficient

E. statements (1) and (2) together are not sufficient, and additional data is needed to answer the question

16. If x and y are both positive integers, how much greater is x than y?

(1) x + y = 20

(2) $x = y^2$

A. statement (1) alone is sufficient, but statement (2) alone is not sufficient to answer the question

B. statement (2) alone is sufficient, but statement (1) alone is not sufficient to answer the question

C. both statements taken together are sufficient to answer the question, but neither statement alone is sufficient

D. each statement alone is sufficient

E. statements (1) and (2) together are not sufficient, and additional data is needed to answer the question

17. Fifty percent of the articles in a certain magazine are written by staff members. Sixty percent of the articles are on current affairs. If 75 percent of the articles on current affairs are written by staff members with more than 5 years experience of journalism, how many of the articles on current affairs are written by staff members with more than 5 years experience?

(1) 20 articles are written by staff members.

(2) Of the articles on topics other than current affairs, 50 percent are by staff members with less than 5 years experience.

A. statement (1) alone is sufficient, but statement (2) alone is not sufficient to answer the question

B. statement (2) alone is sufficient, but statement (1) alone is not sufficient to answer the question

C. both statements taken together are sufficient to answer the question, but neither statement alone is sufficient

D. each statement alone is sufficient

E. statements (1) and (2) together are not sufficient, and additional data is needed to answer the question

18. Is $xy > 0$?

(1) $x/y < 0$

(2) $x + y < 0$

A. statement (1) alone is sufficient, but statement (2) alone is not sufficient to answer the question

B. statement (2) alone is sufficient, but statement (1) alone is not sufficient to answer the question

C. both statements taken together are sufficient to answer the question, but neither statement alone is sufficient

D. each statement alone is sufficient

E. statements (1) and (2) together are not sufficient, and additional data is needed to answer the question

19. One number, n, is selected at random from a set of 10 integers. What is the probability that ½ n + 13 = 0?

(1) The largest integer in the set is 13.

(2) The arithmetic mean of the set is zero.

A. statement (1) alone is sufficient, but statement (2) alone is not sufficient to answer the question

B. statement (2) alone is sufficient, but statement (1) alone is not sufficient to answer the question

C. both statements taken together are sufficient to answer the question, but neither statement alone is sufficient

D. each statement alone is sufficient

E. statements (1) and (2) together are not sufficient, and additional data is needed to answer the question

20. Is w an integer?

(1) 3w is an odd number.

(2) 2w is an even number.

A. statement (1) alone is sufficient, but statement (2) alone is not sufficient to answer the question

B. statement (2) alone is sufficient, but statement (1) alone is not sufficient to answer the question

C. both statements taken together are sufficient to answer the question, but neither statement alone is sufficient

D. each statement alone is sufficient

E. statements (1) and (2) together are not sufficient, and additional data is needed to answer the question

21. A piece of string 1½ meters long is cut into three pieces. What is the length of each of the pieces?

(1) The length of one of the pieces is 20 cm.

(2) The sum of the lengths of two of the pieces is equal to the length of the third piece.

A. statement (1) alone is sufficient, but statement (2) alone is not sufficient to answer the question

B. statement (2) alone is sufficient, but statement (1) alone is not sufficient to answer the question

C. both statements taken together are sufficient to answer the question, but neither statement alone is sufficient

D. each statement alone is sufficient

E. statements (1) and (2) together are not sufficient, and additional data is needed to answer the question

22. There are 120 saplings to be planted in an orchard. Ben and Sue working together without a break can complete the job in six hours. How long would it take Ben working alone to complete the job?

(1) Sue plants 3 saplings in the time it takes Ben to plant 2.

(2) Sue working alone would take 10 hours to do the job.

A. statement (1) alone is sufficient, but statement (2) alone is not sufficient to answer the question

B. statement (2) alone is sufficient, but statement (1) alone is not sufficient to answer the question

C. both statements taken together are sufficient to answer the question, but neither statement alone is sufficient

D. each statement alone is sufficient

E. statements (1) and (2) together are not sufficient, and additional data is needed to answer the question

23. What is the length of the diagonal of rectangle ABCD?

(1) The perimeter of the rectangle is 16.

(2) The area of the rectangle is 16.

A. statement (1) alone is sufficient, but statement (2) alone is not sufficient to answer the question

B. statement (2) alone is sufficient, but statement (1) alone is not sufficient to answer the question

C. both statements taken together are sufficient to answer the question, but neither statement alone is sufficient

D. each statement alone is sufficient

E. statements (1) and (2) together are not sufficient, and additional data is needed to answer the question

24. The arithmetic mean (average) of a set of 10 numbers is 10. Is the median value of the same set also equal to 10?

(1) Exactly half of the numbers are less than 10.

(2) The mode of the set of numbers is 10.

A. statement (1) alone is sufficient, but statement (2) alone is not sufficient to answer the question

B. statement (2) alone is sufficient, but statement (1) alone is not sufficient to answer the question

C. both statements taken together are sufficient to answer the question, but neither statement alone is sufficient

D. each statement alone is sufficient

E. statements (1) and (2) together are not sufficient, and additional data is needed to answer the question

25. When a cookie is taken at random from a jar, what is the probability that it is chocolate flavored?

(1) There are twice as many chocolate flavored cookies as there are almond flavored cookies in the jar.

(2) One third of the cookies in the jar are almond flavored.

A. statement (1) alone is sufficient, but statement (2) alone is not sufficient to answer the question

B. statement (2) alone is sufficient, but statement (1) alone is not sufficient to answer the question

C. both statements taken together are sufficient to answer the question, but neither statement alone is sufficient

D. each statement alone is sufficient

E. statements (1) and (2) together are not sufficient, and additional data is needed to answer the question

26. Yesterday it rained in Baronia. Did it rain in Dukia?

(1) Whenever it rains in Dukia, it also rains in Baronia.

(2) If it does not rain in Dukia, it does not rain in Baronia.

A. statement (1) alone is sufficient, but statement (2) alone is not sufficient to answer the question

B. statement (2) alone is sufficient, but statement (1) alone is not sufficient to answer the question

C. both statements taken together are sufficient to answer the question, but neither statement alone is sufficient

D. each statement alone is sufficient

E. statements (1) and (2) together are not sufficient, and additional data is needed to answer the question

27. How many people are standing in the queue at the counter?

(1) If four more people join the queue, the number in the queue will be more than 15.

(2) If three people give up and leave the queue, the number remaining will be less than 10.

A. statement (1) alone is sufficient, but statement (2) alone is not sufficient to answer the question

B. statement (2) alone is sufficient, but statement (1) alone is not sufficient to answer the question

C. both statements taken together are sufficient to answer the question, but neither statement alone is sufficient

D. each statement alone is sufficient

E. statements (1) and (2) together are not sufficient, and additional data is needed to answer the question

Answer:

1. E

From (1) we know the ratio of male to female sheep, but nothing about the color distribution. So the answer cannot be A or D. From (2) we know that forty percent of the animals are black but nothing about whether they are male of female. So the answer cannot be B. Even putting the information together does not help because there is no way to tell what fraction of the female sheep are black. And so C cannot be correct, and the answer is E.

2. D

From (1) we can tell that a side of E is longer than a side of F, since 3 x side E = 4 x side F. Hence (1) is sufficient to answer the question and the answer must be either A or D. From (2) we could work out the ratio of the lengths of the sides (there is no need to do this, but since we are dealing with regular plane figures the geometry is quite simple), and although we cannot get the actual lengths, we can see from the ratio whether one is bigger than the other. So the answer is D.

3. B

Statement (1) tells us nothing about b and so the answer cannot be A or D. To find what percent a is of b we need to solve the expression (a/b) x 100. Statement (2) allows us to do just that: (a/b) = 1/20. No need to go any further; the answer is B.

4. C

To find the number of revolutions we need to know the rate of turning and the time duration. Statement (1) gives us only the time, and so the answer cannot be A or D. Statement (2) tells us the rate at which a point on the circumference is moving, which, since we know the dimensions of the wheel, is sufficient to determine the number of rotations per minute. But since we do not know the time, B cannot be correct. But putting (1) and (2) together we have all we need, so the answer is C.

5. E

The mean of three numbers will equal the middle number for any set of evenly spaced numbers (1,2,3 or 2,4,6, or -1, -4, -7 for example) and so we cannot assume that x, y and z are consecutive. Hence the answer cannot be A or D. If x, y and z were 2, 4, and 6, for example, the equation in (2) would be valid, so once again the numbers do not have to be consecutive, and the answer cannot be B. From the numbers we have just substituted, we should be able to see that putting (1) and (2) together will still not give a situation in which the numbers are always consecutive or always not consecutive, and so the best answer is E.

6. D

The statement that x is greater than zero means that x is positive. If we multiply a positive number by a negative number the product is negative: this is what we get from (1), which thus tells us that x is positive. The answer must be A or D. The cube of a positive number is positive; the cube of a negative number is negative, and so (2) tells us that x is positive. And so the answer is D.

7. E

It is obvious that neither (1) or (2) alone can tell you how far apart B and C are, and so the answer must be C or E. To see whether putting both pieces of information together will be adequate, visualize two rods: BD of length 10 units, and AC of length 12 units. Mentally place the rods alongside each other so that C lies between B and D. Now you can mentally slide the rods past each other to see that C can lie anywhere between B and D, and so we cannot fix one value for the length BC, and the answer is E.

8. A

The expression $x^2 - y^2$ has factors $(x + y)(x - y)$, and so we can simplify (cancel down) the expression in (1) to get $x + y = 7$. The answer must be A or D. Now consider (2) alone: there are two possible values for $x + y$. (Either 7 or -7). Since there is not one discrete value for (2), the answer must be A.

9. C

From (1) we know the ratio of black socks to white, but that ratio will change when one sock is taken out. To get the new ratio, and hence the probability that the next sock will also be white, we need to know the number of socks of each type. The

answer cannot be A or D. Obviously (2) on its own does not get the ratio and so B cannot be correct. But putting the information in both statements together we can solve the problem (24 white socks with a ratio of black to total of 1:3 means that there are 12 black and 24 white socks). The answer is C.

10. B

Since we need rate of dripping, (1) is not enough and the answer cannot be A or D. Ignoring (1) and looking at (2) we can easily solve the problem because one quarter of the bucket got filled in 2 hours and the filling will get over at 10pm. The answer is B.

11. D

If the price of one ticket is p, and the total number of tickets is n, then from the information given we know that $(6000/n) = p$. From statement (1) we get a new expression $(6000 - 1,200)/n = p - 1$. These two expressions can be solved and so the answer must be A or D. Looking only at (2) we can write another equation: $(8400/n) = p + 2$. This equation can also be solved when combined with the original equation. And so the answer is D.

12. B

To get the ratio we need total numbers male to total numbers female. There is not enough information in (1), so A and D cannot be correct. A ratio is just a fraction, so if we have the fraction of female officers we have the ratio (assuming that all officers have to be either male or female!!!). So the answer is B.

13. B

Statement (1) can have a variety of solutions (n could be 2, or 4 or a fractional number or negative etc. etc.), so the answer cannot be A or D. Looking at (2) on its own we should at least take out the brackets before we decide whether or not it can be solved. So we get $n - m = 2n - 4 - m$ and then we can add m to both sides to give $n = 2n - 4$. In other words we have solved the equation ($n = 4$) and the answer is B.

14. A

Don´t overcomplicate this one! From statement (1) we can directly say that if he had driven at his normal speed he would have taken twice the time (i.e. 70 minutes). So the answer must be A or D. Looking at (2) alone we have speed but not distance and so we cannot get time. The answer is A.

15. D

If a straight line passes through the origin we only need one other point to fix the line. And so we should note that (2) gives us one point and the answer must be B or D. Considering (1) alone, if two lines do not intersect, they must be parallel (have the same slopes). In the equation for a straight line $y = mx + c$, m is the slope. In this case $m = 1$ and we have answered the question. The answer is D.

16. C

Considering (1) alone, there are many pairs of positive integers that fit the equation (19 and 1; 18 and 2 etc.) and so the answer cannot be A or D. Considering (2) alone there are many pairs of numbers that fit the expression (1,1; 4, 2; 16,4 etc.) and so the answer cannot be B. Putting the information together we can solve the problem. First we can get $x = 20-y$ from (1) which we can substitute in the rearranged form of (2): that is $0 = y^2 - x$ to give the quadratic expression $0 = y^2 - 20 + y$. Solving this (no need to bother) will give two solutions but only one will be positive, so the answer is C.

17. A

There is only one item of information in (1), but we can combine it with the given information to find that the total number of articles is 40. Further we can see that 24 are on current affairs, of which 18 are by experienced staff members. So the answer must be A or D. Ignoring (1), we can see that the information in (2) gives us a fraction and not a number and no amount of trying will get us one definite number from which we can find a discrete answer. Hence the answer is A.

18. A

If xy is greater than zero, then either both x and y must be negative, or both must be positive. In statement (1) we have x/y shown as a negative fraction, so either x or y (but not both) must be negative. Hence xy cannot be positive. Since we have a definite solution, the answer must be A or D. Looking at (2) alone, we can see that x and y could both be negative, but it is also possible that one could be negative and the other positive depending on the absolute values of the numbers. Hence the answer is A.

19. E

Statement (1) alone tells us the largest number in the set. If that happened to be the number picked, then we would be certain that ½ n + 13 was not less than or equal to zero. However, we have insufficient information on the other numbers in the set. (If for example -26 was in the set, and the number picked happened to be -26, then we are certain that the value of the expression would be zero.) And so the answer cannot be A or D. From (2) alone we know that some of the numbers must be positive and some negative, but once again, we have no information on the actual numbers. (Do not assume that the smallest number has to be -13). Thus B cannot be the answer. Combining the information is also not sufficient to get the answer because we do not know what the smallest number is. The answer is E.

20. B

If w happened to be 1, then 3w would be an odd number, but we can pick a fraction such as 5/3 for w which also makes 3w an odd number. So the answer cannot be A or D. However, in statement (2) we are told that twice the number is even. All even numbers when divided by 2 will give us whole numbers, and so the answer is B.

21. C

Considering statement (1) we cannot get information on the other two parts of the string. Thus the answer cannot be A or D. From (2) we can write the equation $x + y = z$, and we know that $x + y + z = 150$, but we cannot get any further, and so the answer cannot be B. Now if we take the information in both statements we can see that 20cm cannot be the length of the longest piece (z) and so in our equations we can call it x or y. Now our equations can be solved because we have two equations and two unknowns. ($x + 20 = z$; and $x + 20 + z = 150$. hence $x + 20 + x + 20 = 150$, and $x = 55$). There was no need to solve. Also we should have spotted that since $x + y$ is equal to z, then z is half the string (75cm) and $x + y$ is the other half, so once we have either x or y, we have the solution. Hence answer C.

22. D

From (1) we know how much faster Sue works than Ben does. So we know that of the 120 saplings 48 would be planted by Ben and 72 by Sue. So we know that Ben plants 48/6 = 8 saplings per hour and we can work out how long it would take for 120. Hence the answer must be A or D.

From statement (2) we can get the fraction of the job that Sue does per hour and hence the fraction of the job that she did in six hours. Hence we can work out how much of the job Ben did in six hours, and the fraction he would do per hour. Thus we can get the time it would take him alone. And so the answer is D.

23. C

From (1), 2x + 2y = 16, where x and y are length and breadth. But x and y could take a number of values and so the answer cannot be A or D. From (2) xy = 16, but there are several pairs of numbers that would work (0.5, 32; 1,16; 2,8; 4,4···.), so the answer cannot be B. But if we rearrange the equation we got from (2) as x = 16/y, you should be able to see that we can substitute this in our first equation and solve. (We will get a quadratic with two solutions but one will be negative and we can't have a negative value for a side of a rectangle!) Hence the answer is C.

24. C

From statement (1), we can determine that value of the 5th member of the set is less than 10.

From statement (2), we can determine that value of the 6th member of the set is 10.

Median is (Value 5 + Value 6)/2.

Therefore the median is less than 10. Making the correct answer C.

25. C

To find the probability that the cookie is chocolate flavored, we need to know the number of cookies that are chocolate flavored and the total number of cookies, or we need the ratio of choc flavored to the rest (i.e. the fraction of cookies that are choc flavored). Statement (1) alone is insufficient because there is no way of knowing whether there are other flavors besides choc and almond, and so we don't know what fraction of the total cookies are chocolate. Statement (2) alone does not tell us anything about choc flavored and we cannot assume that two thirds are choc flavored in case there are some other flavors as well. Hence the answer must be C or E. Considering both statements together we can see that if 1/3 are almond, and there are twice as many chocolate, then 2/3 of the cookies must be chocolate. The probability of getting a chocolate cookie is, thus, 2/3, and the answer is C.

26. B

In Dukia it could either rain or not rain. If it rains in Dukia it will also rain in Baronia, but that leaves open the possibility that it could rain in Baronia without it raining in Dukia, and hence statement (1) is not enough to be sure. Now (forgetting statement (1)), if it does not rain in Dukia it will not rain in Baronia, but we know it rained in Baronia, so it could not have been fine in Dukia! So we have answered the question using only statement (2).

27. C

From statement (1) we can deduce that the number in the queue cannot be less than 12. From statement (2) we can deduce that the number in the queue cannot be more than 12. Neither alone gives a definite answer, but taken together there is only one value that fits: 12. Thus the answer is C.

IV. Numerical Reasoning

Numerical Reasoning (數字推理) 基本考核的課題包括度量衡、圖形、分數、數型、單位兌換、組合、排列、方向、時間、數字的分類排列關係結合、因數、倍數、四則運算、價錢、統計、代數、周界面積、算式推斷、基本運算和關係及運算符號等。

試題練習：Numerical Reasoning Q & A

1. In each of the following questions a number series is given with one term missing. Choose the correct alternative that will continue the same pattern and fill in the blank spaces.

   2, 7, 14, 23, ?, 47

   A. 31   B. 28   C. 34   D. 38

2. 4, 6, 12, 14, 28, 30, ?

   A. 32   B. 64   C. 62   D. 60

3. 4, 9, 13, 22, 35, ?

   A. 57   B. 70   C. 63   D. 75

4. 11, 13, 17, 19, 23, 29, 31, 37, 41, ?

   A. 43   B. 47   C. 51   D. 53

5. 15, 31, 63, 127, 255, ?

   A. 513   B. 511   C. 523   D. 517

6. 5, 11, 17, 25, 33, 43, ?

   A. 49   B. 51   C. 52   D. 53

7. 9, 12, 11, 14, 13, ?, 15

   A. 12   B. 16   C. 10   D. 17

8. 0.5, 0.55, 0.65, 0.8, ?

   A. 0.7   B. 0.9   C. 0.95   D. 1

9. 1, 4, 9, 16, 25, ?

   A. 35   B. 36   C. 48   D. 49

10. 2, 1, (1/2), (1/4), ?

   A. (1/3)   B. (1/8)   C. (2/8)   D. (1/16)

11. 1, 4, 27, 16, ?, 36, 343

   A. 125   B. 50   C. 78   D. 132

12. 20, 19, 17, ?, 10, 5

   A. 15   B. 14   C. 13   D. 12

13. 7, 10, 8, 11, 9, 12, ?

   A. 13   B. 12   C. 10   D. 7

14. 6, 11, 21, 36, 56, ?

   A. 51   B. 71   C. 81   D. 41

15. 2, 3, 5, 7, 11, ?, 17

    A. 15   B. 14   C. 13   D. 12

16. 36, 34, 30, 28, 24, ?

    A. 26   B. 23   C. 22   D. 20

17. 13, 35, 57, 79, 911, ?

    A. 1145 B. 1113 C. 1117 D. 1110

18. 6, 11, 21, 36, 56, ?

    A. 65   B. 78   C. 81   D. 97

19. 53, 53, 40, 40, 27, 27, ?

    A. 14   B. 12   C. 16   D. 18

20. 1, 6, 13, 22, 33, ?

    A. 35   B. 46   3. 38   4. 49

ANSWERS

1. C

The given sequence is +5, +7, +9,

ie. 2+ 5 = 7, 7 + 7 = 14, 14 + 9 = 23

Missing Number = 23 + 11 = 34.

2. D

The given sequence is a combination of two series 4, 12, 28, ... and 6, 14, 30, ... The pattern is +8, +16, +32. So, the missing number = (28 + 32) = 60

3. A

Sum of two consecutive numbers of the series gives the next number.

4. A

The series consists of prime numbers.

5. B

Each number is double of the preceding one plus 1.

6. D

The sequence is +6, +6, +8, +8, +10, ...

7. B

Alternatively, 3 is added and one is subtracted.

8. D

The pattern is + 0.05, + 0.10, + 0.15, ...

9. B

The sequence is a series of squares, 12, 22, 32, 42, 52....

10. B

This is a simple division series; each number is one-half of the previous number.

11. A

The series consists of cubes of odd numbers and square of even numbers.

12. B

The Pattern is - 1, - 2, -3, ...

13. C

This is a simple alternating addition and subtraction series. In the first pattern, 3 is added; in the second, 2 is subtracted.

14. C

The pattern is + 5, + 10, + 15, + 20, ...

15. C

The series consists of prime numbers starting from 2.

16. C

This is an alternating number subtraction series. The pattern is -2, -4, -2, ...

17. B

The terms are formed by joining together consecutive odd numbers in order. i.e. 1 and 3, 3 and 5, 5 and 7, 7 and 9, 9 and 11,....

18. C

The pattern is + 5, + 10, + 15, + 20,...

19. A

First, each number is repeated, then 13 is subtracted to arrive at the next number.

20. B

The pattern is +5, +7, +9, +11...

V. Interpretation of Tables and Graphs

Interpretation of Tables and Graphs 的測試題形多變，有
Bar Chart、Line Chart、Pie Chart 及 Table Chart，考生宜
勤加練習。

試題練習：Interpretation of Tables and Graphs Q & A

Bar Chart 1 (Question 1 to 5)

Below is the sales of books (in thousand numbers) from six branches - B1, B2, B3, B4, B5 and B6 of a publishing company in 2000 and 2001:

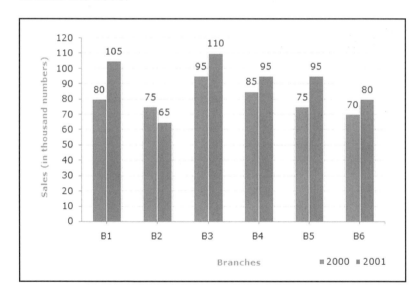

1. What is the ratio of the total sales of branch B2 for both years to the total sales of branch B4 for both years?

   A. 2:3   B. 3:5   C. 4:5   D. 7:9

2. Total sales of branch B6 for both the years is what percent of the total sales of branches B3 for both the years?

A. 68.54% B. 71.11% C. 73.17% D. 75.55%

3. What percent of the average sales of branches B1, B2 and B3 in 2001 is the average sales of branches B1, B3 and B6 in 2000?

A. 75%   B. 77.5%   C. 82.5%   D. 87.5%

4. What is the average sales of all the branches (in thousand numbers) for the year 2000?

A. 73   B. 80   C. 83   D. 88

5. Total sales of branches B1, B3 and B5 together for both the years (in thousand numbers) are?

A. 250   B. 310   C. 435   D. 560

Bar Chart 2 (Question 6 to 10)

The bar graph given below shows the foreign exchange reserves of a country (in million US $) from 1991 - 1992 to 1998 - 1999.

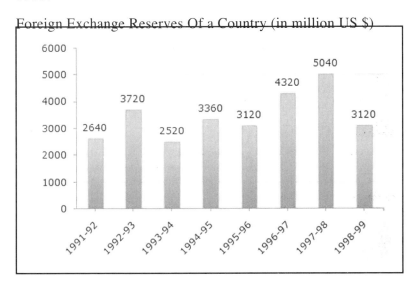

Foreign Exchange Reserves Of a Country (in million US $)

6. The ratio of the number of years, in which the foreign exchange reserves are above the average reserves, to those in which the reserves are below the average reserves is?

    A. 2:6   B. 3:4   C. 3:5   D. 4:4

7. The foreign exchange reserves in 1997-98 was how many times that in 1994-1995?

    A. 0.7   B. 1.2   C. 1.4   D. 1.5

8. For which year, the percent increase of foreign exchange reserves over the previous year, is the highest?

   A. 1992-93 B. 1993-94 C. 1994-95 D. 1996-97

9. The foreign exchange reserves in 1996-97 were approximately what percent of the average foreign exchange reserves over the period under review?

   A. 95%   B. 110%   C. 115%   D. 125%

10. What was the percentage increase in the foreign exchange reserves in 1997-98 over 1993-94?

   A. 100   B. 150   C. 200   D. 620

Bar Chart 3 (Question 11 to 15)

Study the bar chart and answer the question based on it.

Production of Fertilizers by a Company (in 1000 tonnes) Over the Years

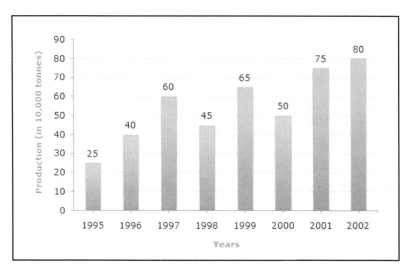

11. What was the percentage decline in the production of fertilizers from 1997 to 1998?

A. 33(1/3)% B. 20%   C. 25%   D. 21%

12. The average production of 1996 and 1997 was exactly equal to the average production of which of the following pairs of years?

A. 2000 and 2001   B. 1999 and 2000

C. 1998 and 2000   D. 1995 and 2001

13. What was the percentage increase in production of fertilizers in 2002 compared to that in 1995?

    A. 320%   B. 300%   C. 220%   D. 200%

14. In which year was the percentage increase in production as compared to the previous year the maximum?

    A. 2002   B. 2001   C. 1997   D. 1996

15. In how many of the given years was the production of fertilizers more than the average production of the given years?

    A. 1   B. 2   C. 3   D. 4

Line Chart 1 (Question 16 to 20)

Study the following line graph and answer the questions.

Exports from Three Companies Over the Years (in Rs. crore)

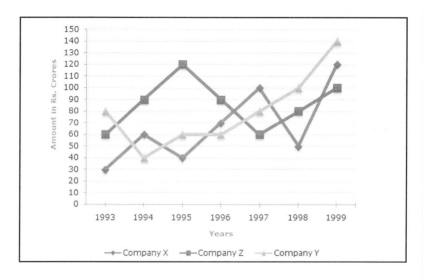

16. For which of the following pairs of years the total exports from the three Companies together are equal?

A. 1995 and 1998   B. 1996 and 1998

C. 1997 and 1998   D. 1995 and 1996

17. Average annual exports during the given period for Company Y is approximately what percent of the average annual exports for Company Z?

A. 87.12% B. 89.64% C. 91.21% D. 93.33%

18. In which year was the difference between the exports from Companies X and Y the minimum?

   A. 1994   B. 1995   C. 1996   D. 1997

19. What was the difference between the average exports of the three Companies in 1993 and the average exports in 1998?

   A. Rs. 15.33 crores   B. Rs. 18.67 crores
   C. Rs. 20 crores   D. Rs. 22.17 crores

20. In how many of the given years, were the exports from Company Z more than the average annual exports over the given years?

   A. 2   B. 3   C. 4   D. 5

Line Chart 2 (Question 21 to 25)

Study the following line graph and answer the questions based on it.

Number of Vehicles Manufactured by Two companies over the Years (Number in Thousands)

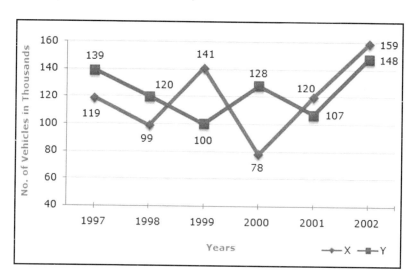

21. What is the difference between the number of vehicles manufactured by Company Y in 2000 and 2001?

A. 50,000   B. 42,000   C. 33,000   D. 21,000

22. What is the difference between the total productions of the two Companies in the given years?

A. 19,000   B. 22,000   C. 26,000   D. 28,000

23. What is the average numbers of vehicles manufactured by Company X over the given period? (rounded off to nearest integer)

A.119,333   B.113,666   C.112,778   D.111,223

24. In which of the following years, the difference between the productions of Companies X and Y was the maximum among the given years?

A. 1997   B. 1998   C. 1999   D. 2000

25. The production of Company Y in 2000 was approximately what percent of the production of Company X in the same year?

A. 173   B. 164   C. 132   D. 97

Line Chart 3 (Question 26 to 30)

The following line graph gives the percent profit earned by two Companies X and Y during the period 1996 - 2001.

Percentage profit earned by Two Companies X and Y over the Given Years

%Profit = [(Income - Expenditure) / Expenditure] x 100

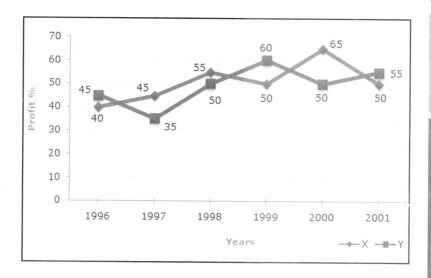

26. The incomes of two Companies X and Y in 2000 were in the ratio of 3:4 respectively. What was the respective ratio of their expenditures in 2000?

    A. 7:22   B. 14:19   C. 15:22   D. 27:35

27. If the expenditure of Company Y in 1997 was Rs. 220 crores, what was its income in 1997?

A. Rs. 312 crores

B. Rs. 297 crores

C. Rs. 283 crores

D. Rs. 275 crores

28. If the expenditures of Company X and Y in 1996 were equal and the total income of the two Companies in 1996 was Rs. 342 crores, what was the total profit of the two Companies together in 1996? (Profit = Income - Expenditure)

A. Rs. 240 crores

B. Rs. 171 crores

C. Rs. 120 crores

D. Rs. 102 crores

29. The expenditure of Company X in the year 1998 was Rs. 200 crores and the income of company X in 1998 was the same as its expenditure in 2001. The income of Company X in 2001 was?

A. Rs. 465 crores

B. Rs. 385 crores

C. Rs. 335 crores

D. Rs. 295 crores

30. If the incomes of two Comapanies were equal in 1999, then what was the ratio of expenditure of Company X to that of Company Y in 1999?

A. 6:5   B. 5:6   C. 11:6   D. 16:15

Pie Chart 1 (Question 31 to 35)

The following pie-chart shows the percentage distribution of the expenditure incurred in publishing a book. Study the pie-chart and the answer the questions based on it.

Various Expenditures (in percentage) Incurred in Publishing a Book

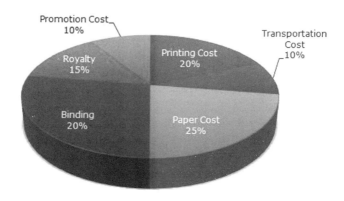

31. If for a certain quantity of books, the publisher has to pay Rs. 30,600 as printing cost, then what will be amount of royalty to be paid for these books?

   A. Rs. 19,450   B. Rs. 21,200
   C. Rs. 22,950   D. Rs. 26,150

32. What is the central angle of the sector corresponding to the expenditure incurred on Royalty?

   A. 15°   B. 24°   C. 54°   D. 48°

33. The price of the book is marked 20% above the C.P. If the marked price of the book is Rs. 180, then what is the cost of the paper used in a single copy of the book?

A. Rs. 36

B. Rs. 37.50

C. Rs. 42

D. Rs. 44.25

34. If 5500 copies are published and the transportation cost on them amounts to Rs. 82500, then what should be the selling price of the book so that the publisher can earn a profit of 25%?

A. Rs. 187.50

B. Rs. 191.50

C. Rs. 175

D. Rs. 180

35. If the difference between the two expenditures are represented by 18° in the pie chart, then these expenditures possibly are

A. Binding Cost and Promotion Cost

B. Paper Cost and Royalty

C. Binding Cost and Printing Cost

D. Paper Cost and Printing Cost

Pie Chart 2 (Question 36 to 38)

The following pie chart give the information about the distribution of weight in the human body according to different kinds of components. Study the pie charts and answer the question.

Distribution of Weight in Human Body

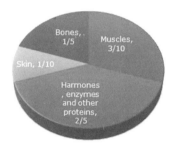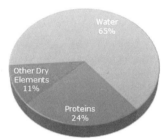

36. What percentage of proteins of the human body is equivalent to the weight of its skin?

   A. 41.66 %   B. 43.33 %

   C. 44.44 %   D. Cannot be determined

37. How much of the human body is neither made of bones or skin?

   A. 40 %   B. 50 %   C. 60 %   D. 70 %

38. What is the ratio of the distribution of proteins in the muscles to that of the distribution of proteins in the bones?

   A. 2:1   B. 2:3   C. 3:2   D. Cannot be determined

Pie Chart 3 (Question 39 to 41)

The pie chart shows the distribution of New York market share by value of different computer companies in 2005.

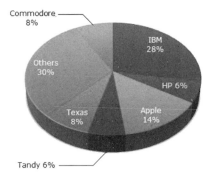

The pie chart shows the distribution of New York market share by volume of different computer companies in 2005.

Number of units sold in 2005 in New York = 1,500

Value of units sold in 2005 in New York = US $1,650,000.

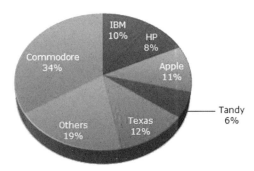

39. For the year 2005, which company has realised the lowest average unit sales price for a PC?

A. Commodore
B. IBM
C. Tandy
D. Cannot be determined

40. Over the period 2005-2006, if sales (value-wise) of IBM PC's increased by 50% and of Apple by 15% assuming that PC sales of all other computer companies remained the same, by what percentage (approximately) would the PC sales in New York (value-wise) increase over the same period?

A. 16.1 %   B. 18 %   C. 14 %   D. None of these

41. In 2005, the average unit sale price of an IBM PC was approximately (in US$)

A. 3,180   B. 2,800   C. 393   D. 3,080

Pie Chart 4 (Question 42 to 46)

The following pie charts show the distribution of students of graduate and post-graduate levels in seven different institutes in a town.

Distribution of students at graduate and post-graduate levels in seven institutes:

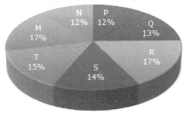

42. What is the total number of graduate and post-graduate level students is institute R?

    A. 8,320   B. 7,916   C. 9,116   D. 8,099

43. What is the ratio between the number of students studying at post-graduate and graduate levels respectively from institute S?

    A. 14:19   B. 19:21   C. 17:21   D. 19:14

44. How many students of institutes of M and S are studying at graduate level?

    A. 7,516   B. 8,463   C. 9,127   D. 9,404

45. What is the ratio between the number of students studying at post-graduate level from institutes S and the number of students studying at graduate level from institute Q?

A. 13:19   B. 21:13   C. 13:8   D. 19:13

46. Total number of students studying at post-graduate level from institutes N and P is

A. 5,601   B. 5,944   C. 6,669   D. 8,372

Table chart 1 (Question 47 to 51)

Study the following table and answer the questions based on it.

Expenditures of a Company (in Lakh Rupees) per Annum Over the given Years.

| Year | Item of Expenditure | | | | |
	Salary	Fuel and Transport	Bonus	Interest on Loans	Taxes
1998	288	98	3.00	23.4	83
1999	342	112	2.52	32.5	108
2000	324	101	3.84	41.6	74
2001	336	133	3.68	36.4	88
2002	420	142	3.96	49.4	98

47. What is the average amount of interest per year which the company had to pay during this period?

    A. Rs. 32.43 lakhs   B. Rs. 33.72 lakhs

    C. Rs. 34.18 lakhs   D. Rs. 36.66 lakhs

48. The total amount of bonus paid by the company during the given period is approximately what percent of the total amount of salary paid during this period?

    A. 0.1%   B. 0.5%   C. 1%   D. 1.25%

49. Total expenditure on all these items in 1998 was approximately what percent of the total expenditure in 2002?

A. 62%   B. 66%   C. 69%   D. 71%

50. The total expenditure of the company over these items during the year 2000 is?

A. Rs. 544.44 lakhs   B. Rs. 501.11 lakhs
C. Rs. 446.46 lakhs   D. Rs. 478.87 lakhs

51. The ratio between the total expenditure on Taxes for all the years and the total expenditure on Fuel and Transport for all the years respectively is approximately?

A. 4:7   B. 10:13   C. 15:18   D. 5:8

Table chart 2 (Question 52 to 56)

The following table gives the percentage of marks obtained by seven students in six different subjects in an examination.

The Numbers in the Brackets give the Maximum Marks in Each Subject.

Student	Subject (Max. Marks)					
	Maths	Chemistry	Physics	Geography	History	Computer Science
	(150)	(130)	(120)	(100)	(60)	(40)
Ayush	90	50	90	60	70	80
Aman	100	80	80	40	80	70
Sajal	90	60	70	70	90	70
Rohit	80	65	80	80	60	60
Muskan	80	65	85	95	50	90
Tanvi	70	75	65	85	40	60
Tarun	65	35	50	77	80	80

52. What are the average marks obtained by all the seven students in Physics? (rounded off to two digit after decimal)

    A. 77.26   B. 89.14   C. 91.37   D. 96.11

53. The number of students who obtained 60% and above marks in all subjects is?

    A. 1   B. 2   C. 3   D. None

54. What was the aggregate of marks obtained by Sajal in all the six subjects?

    A. 409   B. 419   C. 429   D. 449

55. In which subject is the overall percentage the best?

    A. Maths   B. Chemistry
    C. Physics   D. History

56. What is the overall percentage of Tarun?

    A. 52.5%   B. 55%   C. 60%   D. 63%

ANSWERS

1. D

Required ration = (75+65) / (85+95) = 140/180 = 7/9 = 7:9

2. C

Required percentage =[(70+80) / (95+110) x 100]% = (150/205) x 100% = 73.17%

3. D

Average sales (in thousand number) of branches B1, B3 and B6 in 2000 = 1/3 x (80+95+70) = (245/3)

Average sales (in thousand number) of branches B1, B2 and B3 in 2001 = 1/3 x (105+65+110) = 280/3

Therefore, required percentage = [(245/3) / (280/3)] x 100% = (245/280) x 100% = 87.5%

4. B

Average sales of all the six branches (in thousand numbers) for the year 2000 = 1/6 x (80 + 75 + 95 + 85 + 75 + 70) = 80

5. D

Total sales of branches B1, B3 and B5 for both the years (in thousand numbers)

= (80 + 105) + (95 + 110) + (75 + 95)

= 560

6. C

Average foreign exchange reserves over the given period = 3480 million US $.

The country had reserves above 3480 million US $ during the years 1992-93, 1996-97 and 1997-98, i.e., for 3 years and below 3480 million US $ during the years 1991-92, 1993-94, 1994-95, 1995-56 and 1998-99 i.e., for 5 years.

Hence, required ratio = 3:5

7. D

Required ratio = 5040/3360 = 1.5

8. A

There is an increase in foreign exchange reserves during the years 1992 - 1993, 1994 - 1995, 1996 - 1997, 1997 - 1998 as compared to previous year (as shown by bar-graph).

The percentage increases in reserves during these years compared to previous year are:

For 1992-1993 = [(3720 - 2640) / 2640] x 100% = 40.91%

For 1994-1995 = [(3360 - 2520) / 2520] x 100% = 33.33%

For 1996-1997 = [(4320 - 3120) / 3120] x 100% = 38.46%

For 1997-1998 = [(5040 - 4320) / 4320] x 100% = 16.67%

Clearly, the percentage increase over previous year is highest for 1992 - 1993.

9. D

Average foreign exchange reserves over the given period

= 1/8 x (2640 + 3720 + 2520 + 3360 + 3120 + 4320 + 5040 + 3120) million US $

= 3,480 million US $.

Foreign exchange reserves in 1996 - 1997 = 4320 million US $

Therefore, required percentage = (4320/3480) x 100% = 124.14%125%.

10. A

Foreign exchange reserves in 1997 - 1998 = 5040 million US $.

Foreign exchange reserves in 1993 - 1994 = 2520 million US $.

Therefore, increase = (5040 - 2520) = 2520 US $

Therefore, percentage increase = (2520/2520) x 100% = 100%

11. C

Required percentage = [(45-60) / 60] % = -25%

Therefore, there is a decline of 25% in production from 1997 to 1998.

12. D

Average production (in 10,000 tonnes) of 1996 and 1997 = (40+60) / 2 = 50.

We shall find the average production (in 10,000 tonnes) for each of the given alternative pairs:

2000 and 2001 = (50 + 75) / 2 = 62.5

1999 and 2000 = (65 + 50) / 2 = 57.5

1998 and 2000 = (45 + 50) / 2 = 47.5

1995 and 1999 = (25 + 65) / 2 = 45

1995 and 2001 = (25 + 75) / 2 = 50

Therefore, the average production of 1996 and 1997 is equal to the average production of 1995 and 2001.

13. C

Required percentage = [(80-25) / 25] x 100% = 220%

14. D

The percentage increase in production compared to previous year for different years are:

In 1996 = [(40-25) / 25] x 100% = 60%

In 1997 = [(60-40) / 40] x 100% = 50%

In 1998 there is a decrease in production.

In 1999 = [(65-45) / 45] x 100% = 44.44%

In 2000 there is a decrease in production.

In 2001 = [(75-50) / 50] x 100% = 50%

In 2002 = [(80-75) / 75] x 100% = 6.67%

Clearly, there is maximum percentage increase in production in 1996.

15. D

Average production (in 10,000 tonnes) over the given years

= 1/8 (25+40+60 + 45 + 65 + 50 + 75 + 80) = 55

Therefore, the productions during the years 1997, 1999, 2001 and 2002 (total: 4 years) are more than the average production.

16. D

Total exports of the three Companies X, Y and Z together, during various years are:

In 1993 = Rs. (30 + 80 + 60) crores = Rs. 170 crores.

In 1994 = Rs. (60 + 40 + 90) crores = Rs. 190 crores.

In 1995 = Rs. (40 + 60 + 120) crores = Rs. 220 crores.

In 1996 = Rs. (70 + 60 + 90) crores = Rs. 220 crores.

In 1997 = Rs. (100 + 80 + 60) crores = Rs. 240 crores.

In 1998 = Rs. (50 + 100 + 80) crores = Rs. 230 crores.

In 1999 = Rs. (120 + 140 + 100) crores = Rs. 360 crores.

Clearly, the total exports of the three Companies X, Y and Z together are same during the years 1995 and 1996.

17. D

Analysis of the graph: From the graph it is clear that:

1. The amount of exports of Company X (in crore Rs.) in the years 1993, 1994, 1995, 1996, 1997, 1998 and 1999 are 30, 60, 40, 70, 100, 50 and 120 respectively.

2. The amount of exports of Company Y (in crore Rs.) in the years 1993, 1994, 1995, 1996, 1997, 1998 and 1999 are 80, 40, 60, 60, 80, 100 and 140 respectively.

3. The amount of exports of Company Z (in crore Rs.) in the years 1993, 1994, 1995, 1996, 1997, 1998 and 1999 are 60, 90,, 120, 90, 60, 80 and 100 respectively.

Average annual exports (in Rs. crore) of Company Y during the given period

= 1/7 x (80 + 40 + 60 + 60 + 80 + 100 + 140) = 560/7 = 80

Average annual exports (in Rs. crore) of Company Z during the given period

= 1/7 x (60 + 90 + 120 + 90 + 60 + 80 + 100) = 600/7

Therefore, required percentage = [80/(600/7)] x 100%93.33%.

18. C

The difference between the exports from the Companies X and Y during the various years is:

In 1993 = Rs. (80 - 30) crores = Rs. 50 crores.

In 1994 = Rs. (60 - 40) crores = Rs. 20 crores.

In 1995 = Rs. (60 - 40) crores = Rs. 20 crores.

In 1996 = Rs. (70 - 60) crores = Rs. 10 crores.

In 1997 = Rs. (100 - 80) crores = Rs. 20 crores.

In 1998 = Rs. (100 - 50) crores = Rs. 50 crores.

In 1999 = Rs. (140 - 120) crores = Rs. 20 crores.

Clearly, the difference is minimum in the year 1996.

19. C

Average exports of the three Companies X, Y and Z in 1993 = Rs. [1/3 x (30 + 80 + 60)] crores = Rs.(170/3) crores.

Average exports of the three Companies X, Y and Z in 1998 = Rs. [1/3 x (50 + 100 + 80)] crores = Rs.(230/3) crores.

Difference = Rs.[(230/3) − (170/3)] crores = Rs.(60/3) crores = Rs. 20 crores.

20. C

Average annual exports of Company Z during the given period = 1/7 x (60 + 90 + 120 + 90 + 60 + 80 + 100) = Rs. (600/7) crores. = Rs. 85.71 crores

From the analysis of graph the exports of Company Z are more than the average annual exports of Company Z (i.e., Rs. 85.71 crores) during the years 1994, 1995, 1996 and 1999, i.e., during 4 of the given years.

21. D

Required difference = (128,000 − 107,000) = 21000

22. C

From the line-graph it is clear that the productions of Company X in the years 1997, 1998, 1999, 2000, 2001 and 2002 are 119,000, 99,000, 141,000, 78,000, 120,000 and 159,000 and those of Company Y are 139,000, 120,000,100,000, 128,000, 107,000 and 148,000 respectively.

Total production of Company X from 1997 to 2002 = 119,000 + 99,000 + 141,000 + 78,000 + 120,000 + 159,000 = 716,000

and total production of Company Y from 1997 to 2002 = 139,000 + 120,000 + 100,000 + 128,000 + 107,000 + 148,000 = 742,000

Difference = (742,000 − 716,000) = 26,000

23. A

Average number of vehicles manufactured by Company X = 1/6 x (119,000 + 99,000 + 141,000 + 78,000 + 120,000 + 159,000) = 119,333.

24. D

The difference between the productions of Companies X and Y in various years are:

For 1997 (139,000 − 119,000) = 20,000.

For 1998 (120,000 − 99,000) = 21,000.

For 1999 (141,000 − 100,000) = 41,000.

For 2000 (128,000 − 78,000) = 50,000.

For 2001 (120,000 − 107,000) = 13,000.

For 2002 (159,000 − 148,000) = 11,000.

Clearly, maximum difference was in 2000.

25. B

Required percentage = (128,000 / 78,000) x 100% 164%

26. C

Let the incomes in 2000 of Companies X and Y be 3x and 4x respectively.

And let the expenditures in 2000 of Companies X and Y be E1 and E2 respectively.

Then, for Company X we have:

$65 = [(3x - E1) / E1] \times 100$ $65/100 = (3x / E1) - 1$ $E1 = 3x \times (100/165)$ .... (i)

For Company Y we have:

$50 = [(4x - E2) / E2] \times 100$ $50/100 = (4x/ E2) - 1$ $E2 = 4x \times (100/150)$ .... (ii)

From (i) and (ii), we get:

$E1 / E2 = [3x \times (100/165)] / [4x \times (100/150)] = (3 \times 150) / (4 \times 165) = 15/22 = 15:22$ (Required ratio)

27. B

Profit percent of Company Y in 1997 = 35.

Let the income of Company Y in 1997 be Rs. x crores.

Then, $35 = [(x – 220) / 220] \times 100$ $x = 297$

Therefore, income of Company Y in 1997 = Rs. 297 crores.

28. D

Let the expenditures of each companies X and Y in 1996 be Rs. x crores.

And let the income of Company X in 1996 be Rs. z crores.

So that the income of Company Y in 1996 = Rs. (342 - z) crores.

Then, for Company X we have:

$40 = [(z - x) / x] \times 100$ $40/100 = (z / x) -1$ $x = 100z / 140$.... (i)

Also, for Company Y we have:

$45 = [(342 - z) / x] \times 100$ $45/100 = [(342 - z) / x] – 1$ $x = [(342 – z) \times 100] / 145$ ... (ii)

From (i) and (ii), we get:

$100z / 140 = [(342 – z) \times 100] / 145$ $z = 168$ Substituting $z = 168$ in (i), we get : $x = 120$.

Therefore, total expenditure of Companies X and Y in 1996 = 2x = Rs. 240 crores.

Total income of Companies X and Y in 1996 = Rs. 342 crores.

Therefore, total profit = Rs. (342 - 240) crores = Rs. 102 crores.

29. A

Let the income of Company X in 1998 be Rs. x crores.

Then, $55 = [(x – 200) / 200] \times 100$ $x = 310$

Therefore, expenditure of Company X in 2001 = Income of Company X in 1998 = Rs. 310 crores.

Let the income of Company X in 2001 be Rs. z crores.

Then, $50 = [(z – 310) / 310] \times 100$ $z = 465$

Therefore, income of Company X in 2001 = Rs. 465 crores.

30. D

Let the incomes of each of the two Companies X and Y in 1999 be Rs. x.

And let the expenditures of Companies X and Y in 1999 be E1 and E2 respectively.

Then, for Company X we have:

50 = [(x – E1) / E1] x 100 50/100 = (x/E1) – 1   x = (150/100) E1 .... (i)

Also, for Company Y we have:

60 = [(x – E2) / E2] x 100 60/100 = (x/E2) – 1   x = (160/100) E2 .... (ii)

From (i) and (ii), we get:

(150/100) E1 = (160/100) E2   E1/E2 = 160/150 = 16/15 = 16:15 (Required ratio)

31. C

Let the amount of Royalty to be paid for these books be Rs. r.

Then, 20 : 15 = 30600 : r   r = Rs. [(30600 x 15) / 20] = Rs. 22,950

32. C

Central angle corresponding to Royalty = (15% of 360)° = (15/100) x 360° = 54°

33. B

Clearly, marked price of the book = 120% of C.P.

Also, cost of paper = 25% of C.P

Let the cost of paper for a single book be Rs. n.

Then, 120 : 25 = 180 : n n = Rs. (25 x 180) / 120 = Rs. 37.50

34. A

For the publisher to earn a profit of 25%, S.P. = 125% of C.P.

Also Transportation Cost = 10% of C.P.

Let the S.P. of 5,500 books be Rs. x.

Then, 10:125 = 82,500: x Rs. (125 x 82,500) / 10 = Rs. 1,031,250

Therefore, S.P. of one book = Rs. (1,031,250 / 5,500) = Rs. 187.50

35. D

Central angle of 18° =(18/360) x 100% of the total expenditure = 5% of the total expenditure

From the given chart it is clear that:

Out of the given combinations, only in combination (d) the difference is 5%

i.e. Paper Cost - Printing Cost = (25% - 20%) of the total expenditure = 5% of the total expenditure

36. A

Total percentage = (10 / 24) x 100 = 41.6666667% = 41.66%

37. D

20 + 10 = 30% is made up of either bones or skin. Hence, 70% is made up of neither.

38. D

It cannot be determined since the respective distributions are not known.

39. D

Although it seems to be Commodore, the answer cannot be determined due to the fact that we are unaware of the break-up of the sales value and volume of companies compromising the other categories.

40. A

If we assume the total sales to be 100 in the first year, IBM's sales would go up by 50% (from 28 to 42) contributing an increase of 14 to the total sales value.

Similarly, Apple's increase of 15% would contribute an increase of 2.1 to the total sales value. The net change would be 14 + 2.1 on 100. (i.e., 16.1%)

41. D

IBM accounts for 28% of the share by value and 10% of the share by volume.

28% of 1,650,000 = 28 x 1,650,000/100 = 462,000

10% of 1,500 = 10 x 1,500/100 = 150

Therefore, average unit sale price = 462,000/150 = 3,080.

42. D

Required number = (17% of 27,300) + (14% of 24,700)

= 4,641 + 3,458

= 8,099

43. D

Required ratio = (21% of 24,700) / (14% of 27,300) = (21 x 24,700) / (14 x 27,300) = 19/14 = 19:14

44. B

Students of institute M at graduate level= 17% of 27,300 = 4,641.

Students of institute S at graduate level = 14% of 27,300 = 3,822.

Therefore total number of students at graduate in institutes M and S = (4,641 + 3,822) = 8,463.

45. D

Required ratio = (21% of 24700) / (13% of 27300) = (21 x 24,700) / (13 x 27,300) = 19/13 = 19:13

46. C

Required number = (15% of 24,700) + (12% of 24,700) = 3,705 + 2,964 = 6,669

47. D

Average amount of interest paid by the Company during the given period

= Rs. [(23.4 + 32.5 + 41.6 + 36.4 + 49.4)/5] lakhs

= Rs. (183.3 / 5) lakhs

= Rs. 36.66 lakhs

48. C

Required percentage

= [(3.00 + 2.52 + 3.84 + 3.68 + 3.96) / (288 + 342 + 324 + 336 + 420)] x 100%

= (17/1710) x 100% 1%

49. C

Required percentage = [(288 + 98 + 3.00 + 23.4 + 83) / (420 + 142 + 3.96 + 49.4 + 98)] x 100%

= (495.4 / 713.36) x 100%

69.45%

69%

50. A

Total expenditure of the Company during 2000

= Rs. (324 + 101 + 3.84 + 41.6 + 74) lakhs

= Rs. 544.44 lakhs

51. B

Required ratio = [(83 + 108 + 74 + 88 + 98) / (98 + 112 + 101 + 133 + 142)]

= (451 / 586)

= 1/1.3

= 10/13

= 10:13

52. B

Average marks obtained in Physics by all the seven students

= 1/7 x [(90% of 120) + (80% of 120) + (70% of 120) + (80% of 120) + (85% of 120) + (65% of 120) + (50% of 120)]

= 1/7 x [(90 + 80 + 70 + 80 + 85 + 65 + 50)% of 120]

= 1/7 x (520% of 120)

= 624/7

= 89.14

53. B

From the table it is clear that Sajal and Rohit (total 2 students) have 60% or more marks in each of the six subjects.

54. D

Aggregate marks obtained by Sajal

= [(90% of 150) + (60% of 130) + (70% of 120) + (70% of 100) + (90% of 60) + (70% of 40)]

= 135 + 78 + 84 + 70 + 54 + 28

= 449

55. A

We shall find the overall percentage (for all the seven students) with respect to each subject.

The overall percentage for any subject is equal to the average of percentages obtained by all the seven students since the maximum marks for any subject is the same for all the students.

Therefore, overall percentage for:

(i) Maths = [1/7 x (90 + 100 + 90 + 80 + 80 + 70 + 65)]% = [1/7 x 575]% = 82.14%

(ii) Chemistry = [1/7 x (50 + 80 + 60 + 65 + 65 + 75 + 35)]% = [1/7 x 430]% = 61.43%

(iii) Physics = [1/7 x (90 + 80 + 70 + 80 + 85 + 65 + 50)]% = [1/7 x 520]% = 74.29%

(iv) Geography = [1/7 x (60 + 40 + 70 + 80 + 95 + 85 + 77)]% = [1/7 x 507]% = 72.43%

(v) History = [1/7 x (70 + 80 + 90 + 60

+ 50 + 40 + 80)]% = [1/7 x 470]% = 67.14%

(vi) Comp. Science = [1/7 x (80 + 70 + 70 + 60 + 90 + 60 + 80)]% = [1/7 x 510]% = 72.86%

Clearly, this percentage is highest for Maths.

56. C

Aggregate marks obtained by Tarun

= [(65% of 150) + (35% of 130) + (50% of 120) + (77% of 100) + (80% of 60) + (80% of 40)]

= (97.5 + 45.5 + 60 + 77 + 48 + 32)

= 360

The maximum marks (of all the six subjects)

= 150 + 130 + 120 + 100 + 60 + 40

= 600

Therefore, overall percentage of Tarun = (360/600) x 100% = 60%

CHAPTER TWO

模擬試題

Mock Paper 1

演繹推理（8 題）

請根據以下短文的內容，選出一個或一組推論。請假定
短文的內容都是正確的。

1. 田徑場上正在進行100米決賽。參加決賽的是A、B、C、D、E、F六個人。關於誰會得冠軍，看台上甲、乙、丙談了自己的看法：乙認為冠軍不是A就是B。丙堅信冠軍絕不是C。甲則認為D、F都不可能取得冠軍。

   比賽結束後，人們發現他們三個中只有一個人的看法是正確的，請問誰是100米賽冠軍？

   A. A　　　　B. B　　　　C. C　　　　D. E

2. 來自英、法、日、德的甲、乙、丙和丁四位客人，剛好碰在一起。他們除懂本國語言外，每人還會說其他三國語言的一種。有一種語言是三個人都會說的，但沒有一種語言人人都懂，現知道：

   (1) 甲是日本人，丁不會說日語，但他倆都能自由交談。

   (2) 四個人中，沒有一個人既能用日語交談，又能用法語交談。

   (3) 乙、丙、丁交談時，找不到共同語言溝通。

   (4) 乙不會說英語，當甲與丙交談時，他都能做翻譯。

   以下哪項描述正確？（人名後面的文字代表其懂得的語言）

   A. 甲日德、乙法德、丙英法、丁英德
   B. 甲日法、乙日德、丙英法、丁日英
   C. 甲日法、乙法德、丙英德、丁英法
   D. 甲日法、乙英德、丙法德、丁日德

3. 由真正高明的偽造家製造的鈔票從不會被發現，所以一旦他的作品被認出是偽造的，則偽造者不是位高明的偽造者，真正的偽造家從不會被抓到。

下列哪種推理方式與這段話類似？

A. 大衛是一個玩魔術專家，他的魔術總能掩人耳目，從未被揭穿，所以他是一個高明的魔術師。

B. 珍妮是一個玩魔術的人，他的魔術一般不會被揭穿，偶爾有一兩次被人看穿，但這不妨礙他是一名優秀魔術師。

C. 約翰是一個玩魔術的人，他的魔術一般不會被人看穿，偶爾有一兩次被人看穿，說明他並不是一個高明的魔術師，因為高明的魔術師不會被人看穿。

D. 彼德的魔術很好，從不會被揭穿，所以他是一個優秀魔術師。

4. 一家飛機引擎製造商開發出了一種新的引擎，其所具備的安全性能是早期型號的引擎所缺乏的，而早期模型仍然在生產。在這兩種型號的引擎同時被銷售的第一年，早期的型號的銷量超過了新型號的銷量；該製造商於是得出結論認為安全性並非客戶的首要考慮。

下面哪個如果正確，會最嚴重地削弱該製造商的結論？

A. 私人飛機買家和航空公司都從這家飛機引擎製造商那裡購買引擎。

B. 許多客戶認為早期的型號在安全性、風險方面比新型號更小，因為他們對舊型號的安全性知道得更多。

C. 這家飛機引擎製造商的許多客戶也從另一些飛機引擎製造商那裡購買引擎，那些製造商在其新型號引擎中沒有提供額外的安全性能保障。

D. 新型號生產的引擎，可以被所有的使用舊型號引擎的飛機使用。

5. 生命在另外一個行星上發展，必須至少具備兩個條件：(1)適宜的溫度，這是與熱源保持適當距離的結果；(2) 至少在37億年的時間內保持一個相對穩定的溫度變化幅度。這樣的條件在宇宙中很難找到，這使得地球很可能是惟一存在生命的地方。

上述結論成立的前提是：

A. 某一個溫度變化範圍是生命在行星上發展的惟一必要條件

B. 生命不在地球以外的地方生存

C. 在其他行星上的生命形態需要的條件與地球上的生命形態相似

D. 滅絕的生命形態的跡像有可能在有極端溫度的行星上被發現

6. 如果彼德在2000年後從大學畢業，他就必須修讀過世界歷史導論。則這一論點是從下列哪句話中推出？

A. 在2000年前，大學學習中，世界歷史導論不是必修課

B. 每一個選修世界歷史導論的學生都是2000年以後大學畢業的

C. 沒有一個2000年前畢業的大學生修過世界歷史導論

D. 所有2000年後畢業的大學生都必須修世界歷史導論

7. 為了增加收入，一家機場計劃改變其計時停車區收取的停車費。機場會在第一個4小時或不到4小時期間收取4美元，之後每小時收取1美元；而不是在第一個2小時或不到2小時期間收取2美元，之後每小時收取1美元。

下面哪種考慮，如果正確，表明該計劃可以成功地增加收入？

A. 很少有人會在機場的計時停車區內一次停車超過2小時。

B. 在過去的幾年內，機場運營計時停車設備的成本要高於從中獲得的收入。

C. 把車停在機場進行短途旅行的人通常把車停在按天計費而非按時計費的停車區內。

D. 用來運營機場停車區的資金很大一部分被用來維護設備而不是支付收取停車費的職工工資。

8. 一堂考試試卷上畫了五大洲的圖形，每個圖形都編了代號，要求填出其中任意兩個洲名。

甲填：3是歐洲，2是美洲。

乙填：4是亞洲，2是大洋洲。

丙填：1是亞洲，5是非洲。

丁填：4是非洲，3是大洋洲。

戊填：2是歐洲，5是美洲。

結果是他們每人只答對了一半。根據以上條件下列正確的選項是：

A. 1是亞洲，2是歐洲。

B. 2是大洋洲，3是非洲。

C. 3是歐洲，4是非洲。

D. 4是美洲，5是非洲。

II. Verbal Reasoning (English) (6 questions)

In this test, each passage is followed by three statements (the questions). You have to assume what is stated in the passage is true and decide whether the statements are either:

(A) True: The statement is already made or implied in the passage, or follows logically from the passage.

(B) False: The statement contradicts what is said, implied by, or follows logically from the passage.

(C) Can't tell: There is insufficient information in the passage to establish whether the statement is true or false.

Passage 1 (Question 9 to 11)

Alternative investments have become increasingly popular in the last decade as investors seek a safe haven from highly volatile equities and the unpredictable property market. During this time, alternative investments such as art and stamps have outperformed traditional investments by around 60%. High net worth individuals in particular are scrambling to diversify their portfolios in order to mitigate risk and preserve their capital.

While alternative investments can produce higher returns, they have significant downsides. Fakes present a major problem for potential investors as fraudsters flood the market with replicas; this is a particular issue with popular artists such as Picasso, who produced thousands of pieces of art.

Furthermore, art can be expensive to buy and sell, with auctioneers typically taking around 6% of the sale value in fees. Finally, valuable pieces need to insured and protected. The most valuable artworks are often stored in bank vaults and secure art storage facilities, with the owner rarely getting to enjoy the piece.

9. The passage suggests that is safer to spread your savings across a range of different investments.

10. Returns on stamp investments have been 60% higher than equity returns over the last ten years.

11. A piece of art is of more aesthetic value to a potential investor if it can be stored at home.

Passage 2 (Question 12 to 14)

The term "free range" is both a method of animal husbandry and a marketing description for poultry products. The term's interpretation depends on where it is applied. In the United States, where there are calls for a stricter definition, free range broadly refers to poultry raised with access to the outdoors. Farmers are not, however, required to provide chickens with access to grass. Within the European Union, free range denotes poultry allowed to roam in open-air, vegetation-covered runs during daylight hours. Additionally, restrictions regulate stocking density. The term "free range" is often used erroneously to describe "yarded" poultry kept in outdoor pens.

Free-range farming is a relatively recent concept, as prior to the twentieth century all animal husbandry was free range. Farmers relied on grass and sunlight to keep livestock healthy before the discovery of vitamins A and D. While free-range chickens, which are vulnerable to predators, cost more to raise than battery chickens and have harder to collect eggs, the resulting product is of higher quality and commands a higher price. Free-range poultry is widely viewed as humane, however animal rights activists argue that this is a misconception. Debeaking, short life spans A. high flock density are common to both free range and battery farming.

12. Yarded chicken is synonymous with free-range chicken

13. Vitamins A and D were discovered in the twentieth century.

14. Free range poultry is less profitable for farmers to produce than battery chickens.

III. Data Sufficient Test (8 questions)

In this test, you are required to choose a combination of clues to solve a problem

15. It takes 3.5 hours for Mathew to row a distance of X km up the stream.

    Find his speed in still water.

    (1) It takes him 2.5 hours to cover the distance of X km downstream.

    (2) He can cover a distance of 84 km downstream in 6 hours.

    A. statement (1) alone is sufficient, but statement (2) alone is not sufficient to answer the question
    B. statement (2) alone is sufficient, but statement (1) alone is not sufficient to answer the question
    C. both statements taken together are sufficient to answer the question, but neither statement alone is sufficient
    D. each statement alone is sufficient
    E. statements (1) and (2) together are not sufficient, and additional data is needed to answer the question

16. A man mixes two types of glues (X and Y) and sells the mixture of X and Y at the rate of $17 per kg. Find his profit percentage.

(1) The rate of X is $20 per kg.

(2) The rate of Y is $13 per kg.

A. statement (1) alone is sufficient, but statement (2) alone is not sufficient to answer the question

B. statement (2) alone is sufficient, but statement (1) alone is not sufficient to answer the question

C. both statements taken together are sufficient to answer the question, but neither statement alone is sufficient

D. each statement alone is sufficient

E. statements (1) and (2) together are not sufficient, and additional data is needed to answer the question

17. A bag contains 20 copper and 10 brass coins. If 9 of the coins are removed, how many copper coins are left in the box?

(1) Of the removed coins, the ratio of the number of copper coins to that of brass coins is 2:1

(2) Four of the first six coins removed are copper.

A. statement (1) alone is sufficient, but statement (2) alone is not sufficient to answer the question

B. statement (2) alone is sufficient, but statement (1) alone is not sufficient to answer the question

C. both statements taken together are sufficient to answer the question, but neither statement alone is sufficient

D. each statement alone is sufficient

E. statements (1) and (2) together are not sufficient, and additional data is needed to answer the question

18. Maria deposits $10,000 in a bank. What is the annual interest which the bank will pay to Maria?

(1) The interest must be paid once every six months.

(2) The rate of interest is 4%.

A. statement (1) alone is sufficient, but statement (2) alone is not sufficient to answer the question

B. statement (2) alone is sufficient, but statement (1) alone is not sufficient to answer the question

C. both statements taken together are sufficient to answer the question, but neither statement alone is sufficient

D. each statement alone is sufficient

E. statements (1) and (2) together are not sufficient, and additional data is needed to answer the question

19. Given that the length of the side of a square is 1 and that the length of the side is increased by x%. State whether the area of the square is increased by more than 10%.

(1) x < 10

(2) x > 5

A. statement (1) alone is sufficient, but statement (2) alone is not sufficient to answer the question

B. statement (2) alone is sufficient, but statement (1) alone is not sufficient to answer the question

C. both statements taken together are sufficient to answer the question, but neither statement alone is sufficient

D. each statement alone is sufficient

E. statements (1) and (2) together are not sufficient, and additional data is needed to answer the question

20. People in a club either speak French or Russian or both. Find the number of people in a club who speak only French.

(1) There are three hundred people in the club and the number of people who speak both French and Russian is 196.

(2) The number of people who speak only Russian is 58.

A. statement (1) alone is sufficient, but statement (2) alone is not sufficient to answer the question

B. statement (2) alone is sufficient, but statement (1) alone is not sufficient to answer the question

C. both statements taken together are sufficient to answer the question, but neither statement alone is sufficient

D. each statement alone is sufficient

E. statements (1) and (2) together are not sufficient, and additional data is needed to answer the question

21. Joe is older to Lloyd by five years. Ten years ago, John was 10 years older than Mary. What is Mary's age today?

(1) Mary's age today is three times the age of Joe.

(2) Lloyd today is 5 years old.

A. statement (1) alone is sufficient, but statement (2) alone is not sufficient to answer the question

B. statement (2) alone is sufficient, but statement (1) alone is not sufficient to answer the question

C. both statements taken together are sufficient to answer the question, but neither statement alone is sufficient

D. each statement alone is sufficient

E. statements (1) and (2) together are not sufficient, and additional data is needed to answer the question

22. A sum of $385 was divided among Jack, Pollock and Gibbs. Who received the minimum amount?

(1) Jack received 2/9 of what Pollock and Gibbs together received.

(2) Pollock received 3/11 of what Jack and Gibbs together received.

A. statement (1) alone is sufficient, but statement (2) alone is not sufficient to answer the question

B. statement (2) alone is sufficient, but statement (1) alone is not sufficient to answer the question

C. both statements taken together are sufficient to answer the question, but neither statement alone is sufficient

D. each statement alone is sufficient

E. statements (1) and (2) together are not sufficient, and additional data is needed to answer the question.

IV. Numerical Reasoning (5 questions)

Each question is a sequence of numbers with one or two numbers missing. You have to figure out the logical order of the sequence to find out the missing number(s).

23. 2, 7, 14, 23, ?, 47

    A. 31   B. 28   C. 34   D. 38

24. 4, 6, 12, 14, 28, 30, ?

    A. 32   B. 64   C. 62   D. 60

25. 4, 9, 13, 22, 35, ?

    A. 57   B. 70   C. 63   D. 75

26. 11, 13, 17, 19, 23, 29, 31, 37, 41, ?

    A. 43   B. 47   C. 51   D. 53

27. 15, 31, 63, 127, 255, ?

    A. 513   B. 511   C. 523   D. 517

V. Interpretation of Tables and Graphs (8 questions)

Graph 1 (Question 28 to 29)

The bar graph given below shows the data of the production of paper (in lakh tonnes) by three different companies X, Y and Z over the years.

Production of Paper (in lakh tonnes) by Three Companies X, Y and Z over the Years

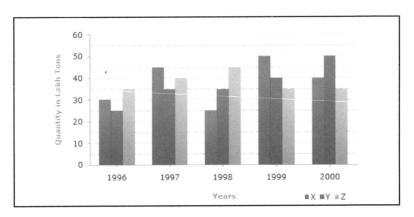

28. For which of the following years, the percentage rise/fall in production from the previous year is the maximum for Company Y?

A. 1997   B. 1998   C. 1999   D. 2000

29. What is the ratio of the average production of Company X in the period 1998-2000 to the average production of Company Y in the same period?

A. 1:1　　B. 15:17　　C. 23:25　　D. 27:29

Graph 2 (Question 30 to 31)

The following line graph gives the ratio of the amounts of imports by a company to the amount of exports from that company over the period from 1995 to 2001.

Ratio of Value of Imports to Exports by a Company Over the Years

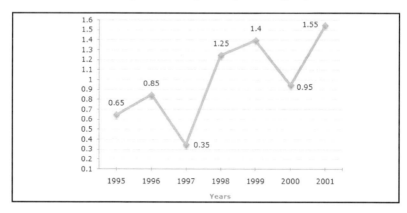

30. If the imports in 1998 was Rs. 250 crores and the total exports in the years 1998 and 1999 together was Rs. 500 crores, then the imports in 1999 was?

A. Rs. 250 crores　　　B. Rs. 300 crores

C. Rs. 357 crores　　　D. Rs. 420 crores

31. The imports were minimum proportionate to the exports of the company in the year?

    A. 1995   B. 1996   C. 1997   D. 2000

Graph 3 (Question 32 to 33)

The following pie charts exhibit the distribution of the overseas tourist traffic from India. The two charts show the tourist distribution by country and the age profiles of the tourists respectively.

Distribution of Overseas Tourist Traffic from India

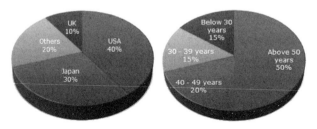

32. The ratio of the number of Indian tourists that went to USA to the number of Indian tourists who were below 30 years of age is?

    A. 2:1   B. 8:3   C. 3:8   D. Cannot be determined

33. If amongst other countries, Switzerland accounted for 25% of the Indian tourist traffic, and it is known from official Swiss records that a total of 25 lakh Indian tourists had gone to Switzerland during the year, then find the number of 30-39 year old Indian tourists who went abroad in that year?

A. 18.75 lakh

B. 25 lakh

C. 50 lakh

D. 75 lakh

Graph 4 (Question 34 to 35)

Study the following table and answer the questions.

Number of Candidates Appeared and Qualified in a Competitive Examination from Different States Over the Years

State	Year									
	1997		1998		1999		2000		2001	
	App.	Qual.	App.	Qual.	App.	Qual.	App.	Qual.	App.	Qual.
M	5200	720	8500	980	7400	850	6800	775	9500	1125
N	7500	840	9200	1050	8450	920	9200	980	8800	1020
P	6400	780	8800	1020	7800	890	8750	1010	9750	1250
Q	8100	950	9500	1240	8700	980	9700	1200	8950	995
R	7800	870	7600	940	9800	1350	7600	945	7990	885

34. Total number of candidates qualified from all the states together in 1997 is approximately what percentage of the total number of candidates qualified from all the states together in 1998?

    A. 72%   B. 77%   C. 80%   D. 83%

35. In which of the given years the number of candidates appeared from State P has maximum percentage of qualified candidates?

    A. 1997   B. 1998   C. 1999   D. 2001

答案

1. C

甲認為A、B、C或E是冠軍，乙認為A或B是冠軍，丙則認為A、B、D、E或F可能是冠軍。由於三人當中只有一個人正確，所以冠軍應該是C或F，但選項中只有C，故答案為C。

2. A

此題可直接用觀察選項法得出正確答案，根據第二條規則，日語和法語不能同時由一個人說，所以B、C、D都錯誤，只有A正確，再將A代入題幹驗證，可知符合條件。

3. C

題幹的推理過程是：高明的偽造家不會被發現偽造，一旦被發現了偽造，即證明該偽造者不是高明的偽造家。C項過程類似：高明的魔法師不會被人看穿，一旦被看穿的話，就說明不是高明的魔法師。答案為C。

4. B

推論的假設是客戶對新、舊款引擎在安全性能方面的了解程度一樣，所以要削弱該推論，只要否定了這個假設即可，答案為B。

5. C

推論的得出需要一個假設，即只有其他星球的生命形態需要的條件和地球上生命形態需要的條件一致，因為在宇宙中難以找到具備兩個必要條件的星球推導出地球可能是惟一存在生命的地方，故答案為C。

6. D

彼德是2000年後從大學畢業的大學生之一，所以，如果這些大學生必須修過世界歷史導論的話，他也一定如此，所以D項是結論的前提。

7. A

按照原來的收費標準，如果不足兩小時，則收費兩元，按新標準則是四元，而在兩小時以上四小時以內、四小時以上，兩者價格變得一樣，所以要增加收入，只能在兩小時以內停車的人多，故A為正確選項。

8. C

此題用假設法。假設甲「3是歐洲」的說法正確，那麼2就不是美洲。同時，2也不是歐洲，5是美洲(由戊所說推出)，再根據丙所說知道1是亞洲，然後根據乙所說得出2是大洋州，最後根據丁的說法知道4是非洲。答案為C。

9. A (True)

The third sentence explains that the very rich have been diversifying their portfolios to reduce the risk of, as the first sentence explains, there being sudden changes in share prices.

10. C (Can't tell)

The passage states that alternative investments such as stamps have "outperformed traditional investments by around 60%". However this statement does specify that equity investments are a type of traditional investment.

11. A (True)

As the last sentence in the passage explains, there is a downside to investing in expensive art that then has to be stored in a bank vault. Part of what is lost is indeed the aesthetic value of being able to appreciate how the art looks.

12. B (False)

The first paragraph states that "The term "free range" is often used erroneously to describe "yarded" poultry kept in outdoor pens." While the term is used synonymously, this is in fact a mistake.

13. C (Can't tell)

Be careful if you answered "true" to this question. It is indeed true that Vitamins A and D were discovered in the twentieth century. However, this was not stated in the passage. It is easy to assume this, based on the fact that the second paragraph states that "Free-range farming is a relatively recent concept, as prior to the twentieth century all animal husbandry was free range" and then goes on to say "Farmers relied on grass and sunlight to keep livestock healthy before the discovery of vitamins A and D". But it is not possible to say definitively when these vitamins were discovered.

14. C (Can't tell)

The second paragraph details both advantages and disadvantages to producing free range chicken. While it costs more to raise free range chickens, they can be sold at a higher price. No direct economic comparison is given with battery chickens.

15. C

Given that Mathew rows upstream with the speed of X / 3.5km/h.

Combining both the statements, we can calculate the downstream speed.

Downstream speed = 84 / 6 = 14 km/h.

Also, downstream speed = X / 2.5.

Or, X / 2.5 = 14.

Or X = 2.5 * 14 = 35 km.

Hence the upstream speed = X / 3.5

= 35 / 3.5

= 10 km/h.

So the speed in still water = (10 + 4) / 2 = 12 km/h.

Hence we need both the statements together to solve the question.

16. E

In order to find the profit or loss, the most important information we need to know is the ratio of X and Y. Neither of the statements provide us with any information regarding the ratios. Both the statements give only the rate of X and Y. Hence the given information is not sufficient to answer the given question.

17. A

Let the number of copper and brass coins removed be 2x and x respectively (from first statement). Now, given that 2x + x = 9 or x = 3.

Hence we can conclude that the number of copper coins removed is 6.

Hence statement (1) alone is sufficient.

Statement (2) gives only half the information. If the information about the other three coins (that were removed) had also been given, it would have been possible for us to find the answer. Hence statement (2) alone is insufficient.

18. B

The only thing we need to know while calculating the annual interest is the rate of interest which is given in only statement (2). Hence it is the only statement which is sufficient.

19. B

According to the first statement, if x < 10, it can be any number between 0 to 9. In such case, the area may or may not increase by more than 10%.

Hence statement (1) alone is insufficient.

Now, if we take x="5," then the area increases by 10.25%. Hence for every value of x > 5, the area has to increase by more than 10%. Hence statement (2) alone is sufficient.

20. C

Let the number of people who speak French be p, that of who speak both be q, and that of who speak only Russian be r.

According to statement (1), p + q + r = 300 and q = 196

But we need the value of r to calculate the value of p. Hence statement (1) alone is insufficient.

Statement (2) gives only the value of r. Hence p and q cannot be found by

statement (2) alone. Hence statement (2) alone is also insufficient.

However, from both the statements, we get, p = 300 - q - r = 300 - 196 -

58 = 46. Thus the value can be found. Hence we need both the

statements together to answer the given question.

21. C

Let the ages of Joe, Lloyd, John and Mary be p, q, r and s.

It is also given that p = q + 5, and r = s + 10. Now, we have to find the

value of s.

According to statement (1), s = 3p. However as their present ages are not

given, hence 1 alone is insufficient to find the answer.

According to statement (2), q = 5. Hence the value of p can be calculated to

be 10. However as statement (2) does not alone give any relation between

p and s, it alone is insufficient to answer the question.

Using both the statements together, we find that s = 3p = 3 * 10 = 30.

Hence we require both the statements together to answer the given question.

22. C

Let the amount received by Jack, Pollock and Gibbs be x, y and z respectively.

Also, x + y + z = 385. ------- (1)

According to statement (1), x = (2/9) (y + z). -------- (2)

This gives us two equations. But there are three unknowns to be found.

Hence statement (1) alone is insufficient.

According to statement (2), y = (3/11) (x + z) ------- (3)

And x + y + z = 385 [equation ---- (1)]

Again we have two equations and three unknowns. Hence statement (2)

alone is also insufficient to find the answer.

However, if we combine both the statements together, we get three

different equations and three unknowns. Hence we need both the

statements to find the answer.

23) C

The given sequence is +5, +7, +9,

ie. 2+ 5 = 7, 7 + 7 = 14, 14 + 9 = 23

Missing Number = 23 + 11 = 34.

24) D

The given sequence is a combination of two series 4, 12, 28, ... and 6, 14, 30, ... The pattern is +8, +16, +32. So, the missing number = (28 + 32) = 60

25) A

Sum of two consecutive numbers of the series gives the next number.

26) A

The series consists of prime numbers.

27) B

Each number is double of the preceding one plus 1.

28. A

Percentage change (rise/fall) in the production of Company Y in comparison to the previous year, for different years are:

For 1997 = [(35-25) / 25] x 100% = 40%

For 1998 = [(35-35) / 25] x 100% = 0%

For 1999 = [(40-35) / 35] x 100% = 14.29%

For 2000 = [(50-40) / 40] x 100% = 25%

Hence, the maximum percentage rise/fall in the production of Company Y is for 1997.

29. C

Average production of Company X in the period 1998-2000

= 1/3 x (25 + 50 + 40) = 115/3 lakh tons

Average production of Company Y in the period 1998-2000

= [1/3 x (35 + 40 + 50)] = 125/3 lakh tons

Therefore, required ratio = [(115/3) / (125/3)] = 115 / 125 = 23/25 = 23:25

30. D

The ratio of imports to exports for the years 1998 and 1999 are 1.25 and 1.40 respectively.

Let the exports in the year 1998 = Rs. x crores.

Then, the exports in the year 1999 = Rs. (500 - x) crores.

Therefore, 1.25 = (250/x)   x = 250 / 1.25 = 200 [Using ratio for 1998]

Thus, the exports in the year 1999 = Rs. (500 - 200) crores = Rs. 300 crores.

Let the imports in the year 1999 = Rs. y crores.

Them, 1.40 = y / 300   y = (300 x 1.40) = 420

Therefore, imports in the year 1999 = Rs. 420 crores.

31. C

The imports are minimum proportionate to the exports implies that the ratio of the value of imports to exports has the minimum value.

Now, this ratio has a minimum value 0.35 in 1997, i.e., the imports are minimum proportionate to the exports in 1997.

32. B

40:15 = 8:3

33. D

Tourist traffic from other countries to Swiz is 20%.

Amongst this 20%, 25% of traffic from India.

So, 25% of 20% = 5% corresponds to the Indian traffic in Switzerland.

5 % corresponds to Switzerland's 25 lakh. Hence 15% will be 75 lakh.

34. C

Required percentage = [(720 + 840 + 780 + 950 + 870) / (980 + 1050 + 1020 + 1240 + 940)] x 100%

= (4160 / 5230) x 100%

= 79.54%

80%

35. D

The percentages of candidates qualified to candidates appeared from State P during different years are:

For 1997: (780/6,400) x 100% = 12.19%

For 1998: (1,020/8,800) x 100% = 11.59%

For 1999: (890/7,800) x 100% = 11.41%

For 2000: (1,010/8,750) x 100% = 11.54%

For 2001: (1,250/9,750) x 100% = 12.82%

Therefore, maximum percentage is for the year 2001.

Mock Paper 2

演繹推理（8 題）

請根據以下短文的內容，選出一個或一組推論。
請假定短文的內容都是正確的。

1. 甲、乙、丙三個人討論一數學題,當她們都把自己的解法說出來以後,甲說:「我做錯了。」乙說:「甲做對了。」丙說:「我做錯了。」老師看過他們的答案並聽了她們的上述意見後說:「你們三個人有一個做對了,有一個説對了」。那麼,誰做對了呢?

   A. 甲　　　B. 乙　　　C. 丙　　　D. 不能確定

2. 要從代號為A、B、C、D、E、F六個偵查員中,挑選幾個人去破案,人選的配備要求必須注意下列各點:

   (1) A、B兩人中至少去一人

   (2) A、D不能一起

   (3) A、E、F三人中要派兩人去

   (4) B、C兩人都去或都不去

   (5) C、D兩人中去一人

   (6) 若D不去、則E也不去

   由此可知:

   A. 挑了A、B、F三人去
   B. 挑了A、B、C、F四人去
   C. 挑了B、C、E三人去
   D. 挑了B、C、D、E四人去

3. 醫學界對 5 種抗菌素進行了藥效比較，得到結果如下：甲藥比乙藥有效，丙藥的毒副作用比丁藥大，戊藥的藥效最差，乙藥與己藥的藥效相同。由此可知：

   A. 甲藥與丁藥的藥效相同
   B. 戊藥的毒副作用最大
   C. 甲藥是最有效的藥物
   D. 己藥比甲藥的藥效差

4. 一定的經濟發展水平，只能支持一定數量和質量的人口，因而物質資料的生產和人口增長必須協調發展。人作為生產者、消費者，其數量和質量必須與生產資料的質與量、消費品的結構與數量，以及資金的數量與投資結構等相適應。由上可以推出：

   A. 目前中國人口數量與其經濟發展水平已不相適應。
   B. 人既是生產者，又是消費者，但生產出的價值遠大於消費掉的。
   C. 提高了人的數量和質量，經濟就會發展。
   D. 當人的增長數量超過經濟發展水平時，人的消費質量就會下降。

5. 在過去的40年內，不僅農業用殺蟲劑的數量大大增加，而且農民們使用殺蟲劑時的精心和熟練程度也不斷增加。然而，在同一時期內，某些害蟲在世界範疇內對農作物造成的損失的比例也上升了，即使在這些害蟲還沒有產生對現有殺蟲劑的抵抗性時也是如此。

下列哪項，如果正確，最好地解釋了為什麼在殺蟲劑使用上的提高伴隨了某些害蟲造成的損失更大？

A. 在40年前通用的一些危險但卻相對無效的殺蟲劑，已經不再在世界範圍內使用了。

B. 由於殺蟲劑對害蟲的單個針對性越來越強，因此，用殺蟲劑來控制某種害蟲的成本在許多情況下，變得比那些害蟲本身造成的農作物損失的價值更大。

C. 由於現在的殺蟲劑對特定使用條件的要求要多於40年前，所以現在的農民們對他們農田觀察的仔細程度要高於40年前。

D. 現在有些農民們使用的某些害蟲控制方法中不使用化學殺蟲劑，但卻和那些使用化學殺蟲劑的害蟲控制方法在減少害蟲方面同樣有效。

6. 教授：在長子繼承權的原則下，男人的第一個妻子生下的第一個男性嬰兒總是首先有繼承家庭財產的權利。

學生：那不正確。侯斯頓夫人是其父惟一妻子的惟一活著的孩子，她繼承了他的所有遺產。

學生誤解了教授的意思，他理解為：

A. 男人可以是孩子的父親

B. 女兒不能算第一個出生的孩子

C. 只有兒子才能繼承財產

D. 私生子不能繼承財產

7. 最受歡迎的電視廣告中有一部分是滑稽廣告，但作為廣告技巧來說，滑稽正是其不利之處。研究表明，雖説很多滑稽廣告的觀眾都能很生動地回憶起這些廣告，但很少有人記得推銷的商品名稱。因此，不管滑稽廣告多麼有趣，多麼賞心悅目，其增加銷售量的能力值得懷疑。上文的假設條件是哪一個？

A. 在觀眾眼裡，滑稽廣告降低了商品信譽。

B. 滑稽廣告雖然可看性強，但常常不如嚴肅的廣告那樣容易被人記住。

C. 不能使商品提高知名度的廣告是不能促進銷售量的增加的。

D. 對滑稽廣告疏遠的觀眾可能和欣賞它的觀眾一樣多。

8. 由於近期的乾旱和高溫，導致海灣鹽度增加，引起了許多魚的死亡。蝦雖然可以適應高鹽度，但鹽度高也給養蝦場帶來了不幸。

以下哪個選項為真，就能夠提供解釋以上現象的原因？

A. 一些魚會游到低鹽度的海域去，來逃脱死亡的厄運。

B. 持續的乾旱會使海灣的水位下降，這已經引起了有關機構的注意。

C. 幼蝦吃的有機物在鹽度高的環境下幾乎難以存活。

D. 水溫生高會使蝦更快地繁殖。

II. Verbal Reasoning (English) (6 questions)

In this test, each passage is followed by three statements (the questions). You have to assume what is stated in the passage is true and decide whether the statements are either:

(A) True: The statement is already made or implied in the passage, or follows logically from the passage.

(B) False: The statement contradicts what is said, implied by, or follows logically from the passage.

(C) Can't tell: There is insufficient information in the passage to establish whether the statement is true or false.

Passage 1 (Question 9 to 11)

In 2008, the mayor of London set a goal of a 400% increase in cycling by 2025. A variety of initiatives have been introduced in order to achieve this target, such as the creation of new bicycle routes into the city, called cycle superhighways.

Based on Paris's popular cycle hire scheme, the Barclays Cycle Hire (BCH) scheme was introduced to London in 2010. Participants pay a small access fee and can then rent bicycles from a fleet of 6,000 and return them to docking stations around the city. Although over 10 million journeys have been taken since the scheme's launch, the BCH is loss-making. The first thirty minutes of any journey are free of charge, so unsurprisingly, 95% of all BCH journeys clock in at under half an hour. The scheme's main income comes from late return penalties.

Cycling in London is widely perceived as dangerous. Cycling advocates believe more measure are needed to unsure London's streets are safe for cyclists. Others argue that the risks are being overstated. Over half a million bike journeys are made in London every day, with the number of cyclists in London increasing by over 80% since the turn of the century. Conversely, cycling fatalities have fallen by 20% since the new millennium.

9. The Barclays Cycle Hire scheme has been unprofitable because cycling in London is seen to be dangerous

10. The majority of participants in the BCH scheme use the bicycles for short journeys

11. Cycling in London is less dangerous today than it was in the 1900s

Passage 2 (Question 12 to 14)

Guano - or bird excrement - has long been a big business in Peru. The rocky islands off the country's Pacific coast are home to large populations of seabirds, such as cormorants, pelicans and boobies. The birds' guano contains high concentrations of phosphorus and nitrogen, making it prized as a natural fertilizer and an ingredient in gunpowder.

Although guano has been valued since the Inca Empire, in the 19th century it became a commodity. By the 1840s, guano represented Peru's main source of income, and was exported to Europe and the United States. Guano extraction was carried out by indentured labours and convicts, many of whom perished. So great was guano's economic importance that it indirectly contributed to several wars. In the early 19th century, however, the guano industry declined following the discovery of nitrogen fixation, in which nitrogen gas is converted into liquid ammonia fertilizer.

Today the Peruvian guano industry is thriving again. Approximately 23,000 tons of guano are sold annually as organic fertilizer. Extraction remains backbreaking manual labour, as machinery frightens birds away. Although poachers kill thousands of birds each year, Peru's seabird population has doubled over the past four years. Overfishing and climate change are the guano's industry's main threats, as seabirds depend on rich anchovy stocks.

12. Guano was the cause of several wars in the 19th century.

13. Guano export formed the backbone of Peru's economy in the 19th century.

14. The guano industry has been revitalized because organic fertilizer is better for the environment than liquid ammonia.

III. Data Sufficient Test (8 questions)

In this test, you are required to choose a combination of clues to solve a problem

15. 'n' is a natural number. State whether $n(n^2 - 1)$ is divisible by 24.

   (1) 3 divides 'n' completely without leaving any remainder.

   (2) 'n' is odd.

   A. statement (1) alone is sufficient, but statement (2) alone is not sufficient to answer the question

   B. statement (2) alone is sufficient, but statement (1) alone is not sufficient to answer the question

   C. both statements taken together are sufficient to answer the question, but neither statement alone is sufficient

   D. each statement alone is sufficient

   E. statements (1) and (2) together are not sufficient, and additional data is needed to answer the question

16. A policeman spots a thief and runs after him. When will the policeman be able to catch the thief?

(1) The speed of the policeman is twice as fast as that of the thief.

(2) The distance between the policeman and the thief is 400 meters.

A. statement (1) alone is sufficient, but statement (2) alone is not sufficient to answer the question

B. statement (2) alone is sufficient, but statement (1) alone is not sufficient to answer the question

C. both statements taken together are sufficient to answer the question, but neither statement alone is sufficient

D. each statement alone is sufficient

E. statements (1) and (2) together are not sufficient, and additional data is needed to answer the question

17. Who got the highest marks among Abdul, Baig and Chiman?

(1) Chiman got half as many marks as Abdul and Baig together got.

(2) Abdul got half as many marks as Baig and Chiman together got.

A. statement (1) alone is sufficient, but statement (2) alone is not sufficient to answer the question

B. statement (2) alone is sufficient, but statement (1) alone is not sufficient to answer the question

C. both statements taken together are sufficient to answer the question, but neither statement alone is sufficient

D. each statement alone is sufficient

E. statements (1) and (2) together are not sufficient, and additional data is needed to answer the question

18. Given that side AC of triangle ABC is 2. Find the length of BC.

(1) AB is not equal to AC

(2) Angle B is 30 degrees.

A. statement (1) alone is sufficient, but statement (2) alone is not sufficient to answer the question

B. statement (2) alone is sufficient, but statement (1) alone is not sufficient to answer the question

C. both statements taken together are sufficient to answer the question, but neither statement alone is sufficient

D. each statement alone is sufficient

E. statements (1) and (2) together are not sufficient, and additional data is needed to answer the question

19. 50% of the people in a certain city have a Personal Computer and an Air conditioner. What percent of people in the city have a personal computer but not an Air-conditioner.

(1) 60% of the people in the city have a Personal Computer.

(2) 70% of the people in the city have an Air-conditioner.

A. statement (1) alone is sufficient, but statement (2) alone is not sufficient to answer the question

B. statement (2) alone is sufficient, but statement (1) alone is not sufficient to answer the question

C. both statements taken together are sufficient to answer the question, but neither statement alone is sufficient

D. each statement alone is sufficient

E. statements (1) and (2) together are not sufficient, and additional data is needed to answer the question

20. Bags I, II and III together have ten balls. If each bag contains at least one ball, how many balls does each bag have?

(1) Bag I contains five balls more than box III.

(2) Bag II contains half as many balls as bag I.

A. statement (1) alone is sufficient, but statement (2) alone is not sufficient to answer the question

B. statement (2) alone is sufficient, but statement (1) alone is not sufficient to answer the question

C. both statements taken together are sufficient to answer the question, but neither statement alone is sufficient

D. each statement alone is sufficient

E. statements (1) and (2) together are not sufficient, and additional data is needed to answer the question

21. Given that $(a + b)^2 = 1$ and $(a - b)^2 = 25$, find the values 'a' and 'b'.

(1) Both 'a' and 'b' are integers.

(2) The value of 'a' = 2

A. statement (1) alone is sufficient, but statement (2) alone is not sufficient to answer the question

B. statement (2) alone is sufficient, but statement (1) alone is not sufficient to answer the question

C. both statements taken together are sufficient to answer the question, but neither statement alone is sufficient

D. each statement alone is sufficient

E. statements (1) and (2) together are not sufficient, and additional data is needed to answer the question

22. Madan is elder than Kamal and Sharad is younger than Arvind. Who among them is the youngest?

(1) Sharad is younger than Madan.

(2) Arvind is younger than Kamal.

A. statement (1) alone is sufficient, but statement (2) alone is not sufficient to answer the question

B. statement (2) alone is sufficient, but statement (1) alone is not sufficient to answer the question

C. both statements taken together are sufficient to answer the question, but neither statement alone is sufficient

D. each statement alone is sufficient

E. statements (1) and (2) together are not sufficient, and additional data is needed to answer the question

IV. Numerical Reasoning (5 questions)

Each question is a sequence of numbers with one or two numbers missing. You have to figure out the logical order of the sequence to find out the missing number(s).

23. 5, 11, 17, 25, 33, 43, ?

    A. 49   B. 51   C. 52   D. 53

24. 9, 12, 11, 14, 13, ?, 15

    A. 12   B. 16   C. 10   D. 17

25. 0.5, 0.55, 0.65, 0.8, ?

    A. 0.7   B. 0.9   C. 0.95   D. 1

26. 1, 4, 9, 16, 25, ?

    A. 35   B. 36   C. 48   D. 49

27. 2, 1, (1/2), (1/4), ?

    A. (1/3)   B. (1/8)   C. (2/8)   D. (1/16)

V. Interpretation of Tables and Graphs (8 questions)

Graph 1 (Question 28 to 29)

Out of the two bar graphs provided below, one shows the amounts (in Lakh Rs.) invested by a Company in purchasing raw materials over the years and the other shows the values (in Lakh Rs.) of finished goods sold by the Company over the years.

Amount invested in Raw Materials (Rs. in Lakhs)

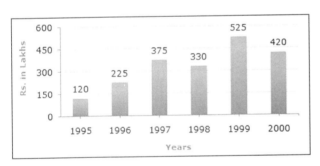

Value of Sales of Finished Goods (Rs. in Lakhs)

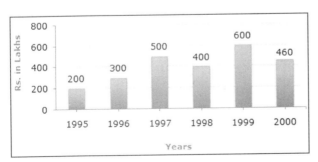

28. In which year, there has been a maximum percentage increase in the amount invested in Raw materials as compared to the year?

A. 1996   B. 1997   C. 1998   D. 1999

29. In which year, the percentage change (compared to the previous year) in the investment on Raw materials is same as that in the value of sales of finished goods?

A. 1996   B. 1997   C. 1998   D. 1999

Graph 2 (Question 30 to 31)

The following line graph gives the percentage of the number of candidates who qualified an examination out of the total number of candidates who appeared for the examination over a period of seven years from 1994 to 2000.

Percentage of Candidates Qualified to Appeared in an Examination Over the Years

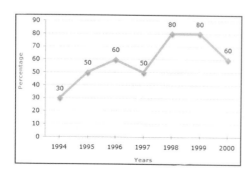

30. The difference between the percentage of candidates qualified to appeared was maximum in which of the following pairs of years?

   A. 1994 and 1995   B. 1997 and 1998
   C. 1998 and 1999   D. 1999 and 2000

31. The total number of candidates qualified in 1999 and 2000 together was 33,500 and the number of candidates appeared in 1999 was 26,500. What was the number of candidates in 2000?

   A. 24,500 B. 22,000 C. 20,500   D. 19,000

Graph 3 (Question 32 to 33)

The following pie-chart shows the sources of funds to be collected by the National Highways Authority of India (NHAI) for its Phase II projects. Study the pie-chart and answers the question that follow.

Sources of funds to be arranged by NHAI for Phase II projects (in crores Rs.)

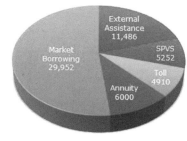

32. If NHAI could receive a total of Rs. 9,695 crores as External Assistance, by what percent (approximately) should it increase the Market Borrowing to arrange for the shortage of funds?

    A. 4.5%   B. 7.5%   C. 6%   D. 8%

33. If the toll is to be collected through an outsourced agency by allowing a maximum 10% commission, how much amount should be permitted to be collected by the outsourced agency, so that the project is supported with Rs. 4,910 crores?

    A. Rs. 6,213 crores   B. Rs. 5,827 crores
    C. Rs. 5,401 crores   D. Rs. 5,316 crores

Graph 4 (Question 34 to 35)

The following table gives the sales of batteries manufactured by a company over the years.

Number of Different Types of Batteries Sold by a Company Over the Years (Numbers in Thousands)

| Year | Types of Batteries | | | | | |
	4AH	7AH	32AH	35AH	55AH	Total
1992	75	144	114	102	108	543
1993	90	126	102	84	126	528
1994	96	114	75	105	135	525
1995	105	90	150	90	75	510
1996	90	75	135	75	90	465
1997	105	60	165	45	120	495
1998	115	85	160	100	145	605

34. The total sales of all the seven years is the maximum for which battery?

   A. 4AH   B. 7AH   C. 32AH   D. 35AH

35. The percentage of 4AH batteries sold to the total number of batteries sold was maximum in the year?

   A. 1994   B. 1995   C. 1996   D. 1997

答案

1. C

此題使用假設法，假設丙做對了，那麼甲、乙都做錯了，這樣，甲説的是正確的，乙、丙説的都錯了，符合條件，答案為C。

2. B

由(3)可以排除C、D，由(4)排除A，因此答案為B，再代入題中驗證，符合條件。

3. D

由於甲藥比乙藥有效，而乙藥又與己藥藥效相同，所以甲藥比己藥藥效好，答案為D。

4. D

一定的經濟發展水平只能支撐一定數量和質量的人口，所以當人的數量過多時，單個個體的消費水平就會下降，答案為D。

5. B

由於殺蟲劑越來越只能針對單個個體發揮作用，所以對於不同的害蟲需要不同的殺蟲劑，結果導致使用殺蟲劑的成本超過了害蟲在沒有被殺死的情況下給農作物造成的損害，而這種成本的增加最終還是應歸結為是害蟲造成的損失，所以答案應該選B。

6. C

根據教授的結論，長子繼承權是特定男性嬰兒的權利，但並不排除女兒也有可能繼承財產，學生忽略了這個可能，所以造成了誤解，在只有女兒的情況下，女兒當然具有繼承財產的權利，這並不會對長子繼承權構成反駁。

7. C

題幹推理過程是：滑稽廣告使人不能記住商品名稱，所以滑稽廣告不能增加銷售量。這裡顯然缺乏一個前提，即商品名稱能否被記住與銷售量的增減有關係。因此答案為C。

8. C

需要解釋的現象是為何蝦能夠適應高鹽度，養蝦場依然會因為鹽度升高而遭遇不幸，所以最恰當的原因就是C項，正因為如此，蝦缺乏食物，所以蝦仍然難以生存。

9. False

While the third paragraph does indeed claim that "Cycling in London is widely perceived as dangerous", the second paragraph states that "10 million journeys have been taken since the scheme's launch". The scheme's economic failure is due to the high percentage of free journeys.

10. True

The second paragraph states that, "95% of all BCH journeys clock in at under half an hour". Hence the majority of journeys are short.

11. Can't Tell

While the last paragraph states that "cycling fatalities have fallen by 20% since the new millennium" the same paragraph also mentions that "Cycling advocates believe more measures are needed to ensure London's streets are safe for cyclists." No direct comparison is given to definitively say whether cycling is more or less dangerous, especially as fatalities are not the only measure of danger – no figures are given for injuries.

12. False

The second paragraph states that guano's economic importance "indirectly contributed" to wars. Therefore it cannot be said to have "caused" the wars.

13. True

The second paragraph states that "By the 1840s, guano represented Peru's main source of income."

14. Can't Tell

There is no comparison given in the passage between liquid ammonia fertilizer and organic fertilizer. You must base your answers only on information provided in the passage.

15. B

According to statement (1), n is a multiple of 3.

Now, say if we take n = 3, the expression is divisible, but in case, we put n

= 6 or 12, then the expression is not divisible by 24. Hence statement (1) alone is insufficient.

Statement (2) alone states that n is odd. Now, if we put any odd value in place of n, we find that the expression is divisible by 24. Hence option 2 alone is sufficient.

16. E

Statement (1) only gives the speeds of both, the thief and the policeman, which cannot be helpful in finding the time. Hence statement (1) alone is insufficient.

Similarly statement (2) gives no clue about the time, it only gives the distance between the two. Hence it alone is also insufficient.

Combining both the statements would also not help us knowing the time.

Hence the answer cannot be found from the given information.

17. C

Let the marks of Abdul, Baig and Chiman be X, Y and Z respectively.

According to statement (1), $Z = \frac{1}{2} (X + Y)$ or $2Z = X + Y$ ------ (1)

We have only one equation but two unknowns. Hence statement (1) alone is insufficient.

According to statement (2), $X = \frac{1}{2} (Y + Z)$ or $2X = Y + Z$ ------ (2)

Again, we have only one equation but two unknowns. Hence statement (2) alone is also insufficient.

However, if we combine both the statements, we get two different equations from which we can find the answer.

18. E

The given properties of the triangle are insufficient to provide any relationship between the sides or the angles of the triangle. It is given that angle Q is 30 degree and side PR is equal to 2. QR could be of any length, which cannot be deduced from the given information.

19. A

According to statement (1), 60 - 50 = 10% of people have a Personal Computer but not an Air-conditioner. Hence statement (1) alone is sufficient to answer the given question.

Statement (2) only helps in finding out what percentage of people have Airconditioner

and not the percentage of people having Personal computer.

Hence it is insufficient to derive the answer.

20. C

From statement (1), only two combinations are possible. Bag III contains 1

and bag I contains 6 or bag III

contains 2 and bag I contains 7 balls. This

information alone is insufficient to answer the given Question.

From statement (2), there are three possibilities; bag II has 1, bag I has 2;

bag II has 2, bag I has 4, and bag II has 3, bag I has 6 balls. Hence it also is insufficient.

If both the statements are combined, we get the possible answer, bag I

has 6, bag III has 1 and bag II has 3 balls. Hence we need both the

statements together to answer the given question.

21. B

On solving both the equations given in the main question, we get ab = - 6. ----- (1)

Now according to statement (1), a and b are integers, they can be [2, - 3]; [- 2, 3]; [1, - 6]; [6, - 1], etc. So statement (1) alone is insufficient.

According to statement (2), a = 2. Hence b = - 2 ---- [from equation ---- (1)]

Hence statement (2) alone is sufficient.

22. B

As given, we have: M > K, A > S.

From II, K > A. Thus, we have: M > K > A > S.

So, Sharad is the youngest. From I, M > S. Thus, we have: M > K > A > S or M > A > K > S or M > A > S > K.

23) D

The sequence is +6, +6, +8, +8, +10, ...

24) B

Alternatively, 3 is added and one is subtracted.

25) D

The pattern is + 0.05, + 0.10, + 0.15, ...

26) B

The sequence is a series of squares, 12, 22, 32, 42, 52....

27) B

This is a simple division series; each number is one-half of the previous number.

28. A

The percentage increase in the amount invested in raw-materials as compared to the previous year, for different years are:

For 1996 = [(225 - 120) /120] % = 87.5%

For 1997 = [(375 - 225) /225] % = 66.67%

For 1998 there is a decrease.

For 1999 = [(525 - 330) /330] % = 59.09%

For 2000 there is a decrease.

Therefore, there is maximum percentage increase in 1996.

29. B

The percentage change in the amount invested in raw-materials and in the value of sales of finished goods for different years are:

Percentage change in Amount invested in raw-material:

For 1996 = [(225 - 120) /120] x 100% = 87.5%

For 1997 = [(375 – 225) / 225] x 100% = 66.67%

For 1998 = [(330 – 375) / 375] x 100% = -12%

For 1999 = [(525 – 330) / 330] x 100% = 59.09%

For 2000 = [(420 - 525) / 525] x 100% = -20%

Percentage change in value of sales of finished goods:

For 1996 = [(300 - 200) /200] x 100% = 50%

For 1997 = [(500 - 300) /300] x 100% = 66.7%

For 1998 = [(400 - 500) /500] x 100% = -20%

For 1999 = [(600 - 400) /400] x 100% = 50%

For 2000 = [(460 - 600) /600] x 100% = -23.33%

Thus, the percentage difference is same during the year 1997.

30. B

The differences between the percentages of candidates qualified to appeared for the give pairs of years are:

For 1994 and 1995 = 50 - 30 = 20

For 1998 and 1999 = 80 - 80 = 0

For 1994 and 1997 = 50 - 30 = 20

For 1997 and 1998 = 80 - 50 = 30

For 1999 and 2000 = 80 - 60 = 20

Thus, the maximum difference is between the years 1997 and 1998.

31. C

The number of candidates qualified in 1999 = (80% of 26,500) = 21,200.

Therefore Number of candidates qualified in 2000 = (33,500 – 21,200) = 12,300.

Let the number of candidates appeared in 2000 be x.

Then, 60% of x = 12,300 x = (12,300 x 100) / 60 = 20,500

32. C

Shortage of funds arranged through External Assistance

= Rs. (11,486 – 9,695) crores

= Rs. 1,791 crores

Therefore, increase required in Market Borrowing = Rs. 1,791 crores

Percentage increase required = (1,791 / 29,952) x 100% = 5.98% 6%

33. C

Amount permitted = (Funds required from Toll for projects of Phase II) + (10% of these funds)

= Rs. 4,910 crores + Rs. (10% of 4,910) crores

= Rs. (4,910 + 491) crores

= Rs. 5,401 crores.

34. C

The total sales (in thousands) of all the seven years for various batteries are:

For 4AH = 75 + 90 + 96 + 105 + 90 + 105 + 115 = 676

For 7AH = 144 + 126 + 114 + 90 + 75 + 60 + 85 = 694

For 32AH = 114 + 102 + 75 + 150 + 135 + 165 + 160 = 901

For 35AH = 102 + 84 + 105 + 90 + 75 + 45 + 100 = 601

For 55AH = 108 + 126 + 135 + 75 + 90 + 120 + 145 = 799.

Clearly, sales are maximum in case of 32AH batteries.

35. D

The percentages of sales of 4AH batteries to the total sales in different years are:

For 1992 = (75/543) x 100% = 13.81%

For 1993 = (90/528) x 100% = 17.05%

For 1994 = (96/525) x 100% = 18.29%

For 1995 = (105/510) x 100% = 20.59%

For 1996 = (96/465) x 100% = 19.35%

For 1997 = (105/495) x 100% = 21.21%

For 1998 = (115/605) x 100% = 19.01%

Clearly, the percentage is maximum in 1997.

CHAPTER THREE

常見問題

什麼人符合申請資格？

- 持有大學學位；

- 現正就讀學士學位課程最後一年；或

- 持有符合申請學位或專業程度公務員職位所需的專業資格。

「綜合招聘考試」(CRE)跟「聯合招聘考試」(JRE)有何分別？

在CRE中英文運用考試中取得「二級」成績後，可投考JRE，考試為AO、EO及勞工事務主任、貿易主任四職系的招聘而設。

CRE成績何時公佈？

考試邀請信會於考前12天以電郵通知，成績會在試後1個月內郵寄到考生地址。

報考CRE的費用是多少？

不設收費。

看得喜 放不低

創出喜閱新思維

書名	投考公務員 能力傾向測試 解題天書 第三版
ISBN	978-988-74806-7-9
定價	HK$118
出版日期	2021年2月
作者	Fong Sir
責任編輯	Y.T.
版面設計	梁文俊
出版	文化會社有限公司
電郵	editor@culturecross.com
網址	www.culturecross.com
發行	香港聯合書刊物流有限公司
	地址：香港新界大埔汀麗路36號中華商務印刷大廈3樓
	電話：（852）2150 2100
	傳真：（852）2407 3062

網上購買 請登入以下網址：

一本 My Book One　　　超閱網 Superbookcity　　　香港書城 Hong Kong Book City

www.mybookone.com.hk　　www.mybookone.com.hk　　www.hkbookcity.com